Passive and Active RF-Microwave Circuits

Series Editor
Pierre-Noël Favennec

Passive and Active RF-Microwave Circuits

Course and Exercises with Solutions

Pierre Jarry
Jacques N. Beneat

ELSEVIER

First published 2015 in Great Britain and the United States by ISTE Press Ltd and Elsevier Ltd

ISTE Press Ltd
27-37 St George's Road
London SW19 4EU
UK

www.iste.co.uk

Elsevier Ltd
The Boulevard, Langford Lane
Kidlington, Oxford, OX5 1GB
UK

www.elsevier.com

Notices
Knowledge and best practice in this field are constantly changing. As new research and experience broaden our understanding, changes in research methods, professional practices, or medical treatment may become necessary.

Practitioners and researchers must always rely on their own experience and knowledge in evaluating and using any information, methods, compounds, or experiments described herein. In using such information or methods they should be mindful of their own safety and the safety of others, including parties for whom they have a professional responsibility.

To the fullest extent of the law, neither the Publisher nor the authors, contributors, or editors, assume any liability for any injury and/or damage to persons or property as a matter of products liability, negligence or otherwise, or from any use or operation of any methods, products, instructions, or ideas contained in the material herein.

For information on all Elsevier publications visit our website at
http://store.elsevier.com/

British Library Cataloguing in Publication Data
A CIP record for this book is available from the British Library
Library of Congress Cataloging in Publication Data
A catalog record for this book is available from the Library of Congress
ISBN 978-1-78548-006-5

Printed and bound in the UK and US

Contents

ACKNOWLEDGMENTS xiii

PREFACE xv

PART 1. MICROWAVE COUPLERS 1

CHAPTER 1. MICROWAVE COUPLED LINES 3

1.1. Introduction 3
1.2. Description 3
1.2.1. Proximity coupler 4
1.2.2. Hybrid ring 5
1.2.3. Guide coupler 6
1.2.4. Use of couplers 7
1.3. Lossless equivalent circuit 7
1.3.1. Mesh law 7
1.3.2. Notch law 8
1.3.3. Coupled differential system of
the first order 9
1.3.4. Non-coupled differential system
of the second order 10
1.4. Homogeneous medium of
permittivity ε 11
1.5. Sinusoidal excitation, even and
odd modes 12
1.6. Bibliography 14

CHAPTER 2. STRIP COUPLER   17

2.1. Introduction   17
2.2. Two coupled lines closed on $R_C(Z_C)$.   18
2.3. Reflection (r) and transmission (t)
of the even and odd modes   19
    2.3.1. The reflection (ρ)   19
    2.3.2. The transmission (t)   20
    2.3.3. Reflection and transmission in the
    even and odd cases   20
2.4. Waves of the result state.   22
2.5. Coupler versus coupling coefficient   25
2.6. Energy   25
2.7. Enlargement of the bandwidth   29
2.8. Scattering matrix   29
2.9. Bibliography.   30

CHAPTER 3. HYBRID AND MAGIC T   33

3.1. Introduction   33
3.2. Coupler hybrid T   34
3.3. Coupler magic T   35
3.4. Application to the determination
of a reflection.   37
    3.4.1. Perfect magic T.   37
    3.4.2. Non-perfect magic T   39
3.5. Bibliography.   42

CHAPTER 4. PROBLEMS.   45

4.1. Practical determination of the
elements of a coupler   45
4.2. Two-stages coupler with length $\theta = \pi/2$ 47
4.3. Perfect directive coupler.   53
4.4. Bibliography.   57

PART 2. MICROWAVE FILTERS   59

CHAPTER 5. ANALYSIS OF A GUIDE RESONATOR
WITH DIRECT COUPLINGS   61

5.1. Introduction   61

5.2. Circuit analysis of the iris alone 62
    5.2.1. Scattering matrix (S). 62
    5.2.2. Chain wave matrix (C) 63
    5.2.3. Tee equivalent circuit 64
5.3. Circuit analysis of the cavity alone 65
5.4. Circuit analysis of a cavity between
two irises 65
    5.4.1. Chain wave matrix (C) 65
    5.4.2. Resonance. 66
    5.4.3. The iris at a frequency near
    the resonance. 67
5.5. Bibliography. 70

CHAPTER 6. ELECTROMAGNETIC (EM) OF THE IRIS 73

6.1. Introduction 73
6.2. Characterization of the iris 73
6.3. Properties of the TE_{m0} modes 75
    6.3.1. Propagating modes ($m = 1$). 75
    6.3.2. Evanescent modes (m $\geq$ 2) 76
    6.3.3. Transport power 77
6.4. Continuities of the waves 77
6.5. Computing the susceptance 79
6.6. Bibliography. 81

CHAPTER 7. SYNTHESIS OF GUIDE FILTERS
WITH DIRECT COUPLING 83

7.1. What does synthesis mean? 83
7.2. Description. 83
7.3. The realizations of the iris. 83
7.4. Synthesis method of the filter. 85
    7.4.1. Butterworth response 87
    7.4.2. Tchebycheff response 87
    7.4.3. Pass-band response 88
7.5. Cavity simulation 89
7.6. Coupling iris simulation 91
7.7. Lengths of the cavities 93
7.8. Practical computing of the filter 94
    7.8.1. Band ratio 94

7.8.2. Invertors 94
7.8.3. Reactances 94
7.8.4. Lengths of the cavities 94
7.9. Capacitive gap filters. 95
7.10. Bibliography 95

CHAPTER 8. PROBLEMS. 97

8.1. Network formed by identical two-port
networks and separated by a guide 97
8.2. Synthesis of a guide filter with direct couplings. 102
8.3. Filters using coupled lines: synthesis of S.B. Cohn 105
8.4. Filters using coupled lines: synthesis of G.L. Matthaei. 109
8.5. Bibliography. 112

PART 3. MICROWAVE AMPLIFIERS. 115

CHAPTER 9. MICROWAVE FET
AMPLIFIERS AND GAINS 117

9.1. Introduction 117
9.2. Recall on the S parameters 118
9.2.1. The network is sourced and loaded by
impedances different from Z_0 120
9.2.2. Flow graph of the load 120
9.2.3. Flow graph of the two-port network 120
9.2.4. Cascade of two two-port networks 121
9.2.5. A load of a two-port network 121
9.2.6. A source. 121
9.3. Masson's rules for non-touching loops 122
9.4. Transducer power gain of a network
with a load and source. 123
9.5. Unilateral transducer gain. 125
9.6. Circles with constant gain
(unilateral case $S_{12} = 0$) 126
9.7. Bibliography. 128

CHAPTER 10. STABILITY 131

10.1. Introduction 131
10.2. Unconditional and conditional stabilities 131

10.3. Limits of stability   133
10.4. Places of stability   136
10.5. Power adaptation in the case of an unconditional stability . . .   137
10.6. Bibliography   142

CHAPTER 11. NOISE   143

11.1. Introduction   143
11.2. Sources of noise   143
11.3. Noise factor   147
11.4. Noise circles   149
11.5. Bibliography   150

CHAPTER 12. PROBLEMS   153

12.1. Symmetric writing of G_T in the
case of the non-unilateral amplifier   153
12.2. Stability conditions of a
broadband transistor from 300 to 900 MHz   157
12.3. Narrow band amplifier around 500 MHz   159
12.4. Low-noise amplifier at 2.5 GHz   167
12.5. Bibliography   172

PART 4. MICROWAVE OSCILLATORS   175

CHAPTER 13. QUASI-STATIC ANALYSIS
AND OVERVOLTAGE COEFFICIENTS
OF AN OSCILLATOR   177

13.1. Introduction   177
13.2. Quasi-static analysis of the
microwave oscillators   178
13.3. NL resistances   179
13.4. Output power of the oscillator   181
13.5. Stability of the oscillation   183
13.6. Overvoltage coefficients of a
microwave oscillator   187
13.6.1. The case of a linear circuit   187
13.6.2. Overvoltage coefficients of an NL circuit   190
13.7. Bibliography   191

CHAPTER 14. SYNCHRONIZATION,
PULLING AND SPECTRA   193

14.1. Introduction   193
14.2. Synchronization   193
14.3. Pulling factor.   198
    14.3.1. Definition   198
    14.3.2. Load variation.   199
14.4. The spectrum of the oscillator.   201
    14.4.1. Frequency modulation noise   203
    14.4.2. Amplitude modulation noise   204
14.5. Bibliography   204

CHAPTER 15. INTEGRATED AND STABLE
MICROWAVE OSCILLATORS USING DIELECTRIC
RESONATORS AND TRANSISTORS   207

15.1. Introduction   207
15.2. A DR coupled to a microstrip line   208
    15.2.1. The scattering matrix   208
    15.2.2. The interpretation.   211
    15.2.3. The influence of the length l
    on the transmission line.   212
15.3. Realization of a stable oscillator
with a DR   214
    15.3.1. Configuration   214
    15.3.2. Characterization of the transistor   215
    15.3.3. Determination of the source
    impedance Z_3   215
    15.3.4. Determination of the gate
    impedance Z_1   220
    15.3.5. Determination of the load
    impedance Z_L   221
15.4. Bibliography   222

CHAPTER 16. PROBLEMS   225

16.1. Scattering parameters
of a transistor   225
16.2. Scattering parameters
and oscillations conditions   228

16.3. Synchronization of an oscillator 233
16.4. Pulling factor of an oscillator 239
16.5. Equivalent circuit of a DR coupled to a line 244
16.6. Bibliography 248

INDEX. 251

Acknowledgments

These courses were given at the universities of Limoges, Brest and Bordeaux (all in France), but also in USA, UK and Brazil.

Pierre Jarry gave also these courses in engineers schools such as Evry (INT), Brest (ENSTBr), Rennes (INSA) and Bordeaux (ENSEIRB).

Pierre Jarry wishes to thank his colleagues at the University of Bordeaux and in particular Professor Eric Kerherve, specialist in microwave amplifiers.

He would like to express his deep appreciation to his wife and his son for their tolerance and support.

Jacques Beneat is very grateful to Norwich University in USA, a place conducive to trying and succeeding in new endeavors.

Finally, we would like to express our sincere appreciation to all the staff at ISTE involved in this project for their professionalism and outstanding efforts.

Preface

Microwave and Radio Frequency (RF) circuits play an important role in communication systems, and due to the proliferation of radar, satellite and mobile wireless systems, there is a need for design methods that can satisfy the ever-increasing demand for accuracy, reliability and fast development times. This book provides basic design techniques for passive and active circuits in the microwave and RF range. It has grown out of the authors' own teaching and as such has a unity of methodology and style, essential for a smooth reading.

The book is intended for microwave engineers and advanced graduate students.

Each of the 16 chapters provides a complete analysis and modeling of the microwave structure used for emission or reception technology. We hope that this will provide students with a set of approaches that they could use for current and future RF and microwave circuit designs. We also emphasize the practical nature of the subject by summarizing the design steps and giving numerous examples of realizations and measured responses so that RF and microwave students can have an appreciation of each circuit. This approach, we believe, has produced a coherent, practical and real-life treatment of the subject. The book is therefore not only theoretical but also experimental with over 16 microwave circuit realizations (couplers, filters, amplifier and oscillators). Problems and exercises constitute about 30% of the book.

Then if we consider, for example, the diagram of an earth station in the C band (Figure P.1), we see that we have at the reception and the emission several microwave elements which can be active or passive:

– active as the amplifiers and the oscillators;

– passive as the different filters and the couplers (couplers are elements that allow us to take a part of the signal in the oscillators).

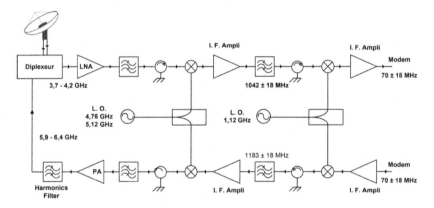

Figure P.1. *Diagram of an earth station in the C band*

We decided to successively study the principal elements that allow the reception and emission of a signal in the cases of earth stations, satellites and RF (mobile phones):

– couplers;

– filters;

– amplifiers;

– oscillators.

For all these four elements (two are passive and two are active), we give their principal properties in three chapters and add one chapter of exercises and problems.

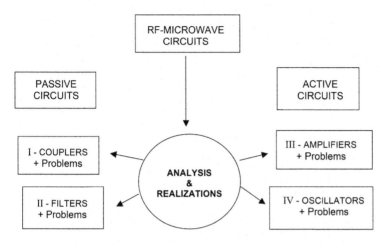

Figure P.2. *Organization of the book*

The book is divided into four parts and 16 chapters.

Part 1 is devoted entirely to the theory and realizations of couplers.

In Chapter 1, we recall the fundamental knowledge about microwave coupled lines and define in particular the even and odd modes.

Chapter 2 gives the analysis of a strip coupler while computing reflection and transmission in the even and odd cases. We define the coupling coefficients, the energy and enlargement of the bandwidth.

Chapter 3 presents the hybrid and magic T and their applications. We consider the cases of the perfect and non-perfect magic T.

In Chapter 4, we present three problems with their solutions. The first problem is devoted to the determination of the elements of a coupler while with the second we compute a two-stage coupler and then we consider (third problem) what happens for a perfect directive coupler. For all three problems, we provide detailed solutions.

In Part 2, we consider how to compute and realize microwave and RF filters.

The goal of Chapter 5 is to design a filter made up of only one resonator (a cavity) and two irises, and we give the response of this very simple filter (out of the resonance and near the resonance).

Chapter 6 deals with the electromagnetic of the iris while considering propagating and evanescent modes.

In Chapter 7, we describe the different methods of synthesis (Tchebycheff and Butterworth) and the possible realizations of the iris. After simulations of the cavities and the iris, we present a method of realizing the entire filter.

The last chapter of the second part (Chapter 8) is devoted to four exercises. The first problem is theoretical because we consider a network composed of 2 two–ports separated by a guide. The second problem is classical and gives the synthesis of a filter made up of a guide with direct couplings. The third and fourth problems give, respectively, the solutions of Cohn and Matthaei.

In Part 3, we consider how to realize microwave and RF amplifiers.

In this case, we give the different flow graphs (Chapter 9) and define, using Masson's rules, the different gains of an amplifier made up of a field effect transistor (FET).

Chapter 10 is devoted to the problems of stability, i.e. the limits of stability and the conditional and unconditional stabilities.

The problem of noise is also very important and it will be given in Chapter 11. In this chapter, we define the sources of noise and also the noise factor and the circles of noise necessary for realizing an amplifier.

In Chapter 12, we present four problems with their solutions. The first problem is devoted to a symmetric of the gain of a non-unilateral amplifier. In the second exercise, we determine the stability conditions of a broadband transistor from 300 to 900 MHz. In the third and fourth exercises, we give the input and output matching circuits for a narrowband amplifier around 500 MHz and for a low noise amplifier around 2.5 GHz.

In the last part (Part 4), we give how to compute and realize microwave and RF oscillators.

First, we have to give (Chapter 13) the quasi-static analysis and the overvoltage coefficients and show that it is possible to extend these definitions to nonlinear circuits as the oscillators.

In Chapter 14, we discuss synchronization and the variation of the frequency with the variation of the load (pulling). We give also the specter of the oscillator.

Chapter 15 is devoted to the realization(s) of oscillators with a very stable element: the dielectric resonator (DR). We consider the coupling of a DR with a microstrip line and the different uses of the active element (FET in our case).

The last chapter (Chapter 16) presents five exercises with their solutions. With the first two exercises, we show how to compute the scattering parameters of a transistor alone and what happens when there are also oscillation conditions. With the third and fourth exercises, we talk of synchronization and pulling. With the last exercise (exercise 5), we give and compute the equivalent circuit of a DR coupled with a line.

These courses and the corresponding problems are given during the fifth year of university and at specialist engineering schools.

Pierre JARRY
France

Jacques N. BENEAT
USA
January 2014

Microwave Couplers

Microwave Coupled Lines

1.1. Introduction

We consider the lines of propagation in which length and coupling are about the wavelength λ. We use to say that the effect is distributed. The electromagnetic (EM) field of the line (1) induces an EM field of the line (2) and reciprocally.

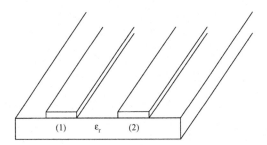

Figure 1.1. *Two coupling microstrip lines*

1.2. Description

The phenomenon is induced on a wavelength λ that is around some *Giga Hertz* $= GHz = 10^9$ Hz in microwaves.

$$\lambda = \frac{c}{f} = \frac{300,000 \text{ km/s}}{10^9} = \frac{3 \times 10^8}{10^9} = 0.3 \text{ m}$$

1 GHz ➜ $\lambda = 30$ cm	decimeter waves
10 GHz ➜ $\lambda = 3$ cm	centimeter waves
100 GHz ➜ $\lambda = 3$ mm	millimeter waves

1.2.1. *Proximity coupler*

We like to say that going from 1 to 2 is the direct way and going from 3 to 4 is the coupling way. Then, we define:

– coupling $\qquad : C(db) = 20 \log \dfrac{1}{|S_{13}|}$

– insertion losses $\qquad : L(db) = 20 \log \dfrac{1}{|S_{12}|}$

– isolation $\qquad : I(db) = 20 \log \dfrac{1}{|S_{14}|}$

– adaptation $\qquad$: Voltage Standing Waves Ratio (VSWR) of
all the $\qquad\qquad\qquad$ lines

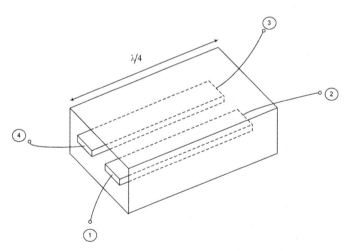

Figure 1.2. *Proximity coupler*

The same definitions occur to the other type of coupler as the hybrid ring or the guide coupler.

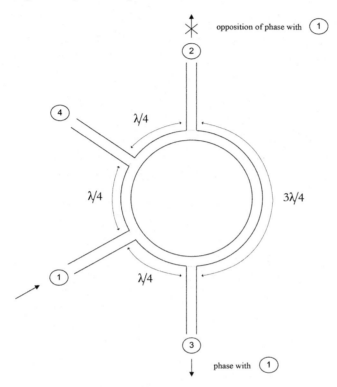

Figure 1.3. *Hybrid ring*

1.2.2. *Hybrid ring*

Suppose that 1 is the input of the waves. From 1 to 3, there is a path of $\lambda/4$. Through the other way, 1, 4, 2, 3, there is a path of $5\lambda/4$. And a difference of way of:

$$\Delta l = \frac{5\lambda}{4} - \frac{\lambda}{4} = \lambda$$

Then, 3 is in phase with 1 and this recombines two waves in 3.

Now, consider 1 as the input and 2 as the output. The way 1, 3, 2 is $4\lambda/4$ and the way 1, 4, 2 is $2\lambda/4$. This induces a difference in way of:

$$\Delta l = \frac{4\lambda}{4} - \frac{2\lambda}{4} = \frac{\lambda}{2}$$

And nothing appears in 2. We have constructed a coupler.

1.2.3. Guide coupler

Coupling is made by the two irises I_1 and I_2 upon a length of $\lambda_g/4$.

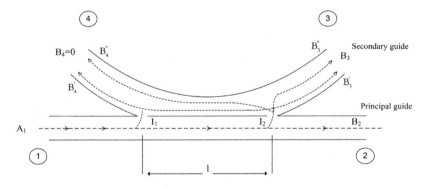

Figure 1.4. *Guide coupler*

On iris I_1, a small part of the energy (10^{-3}) goes through the secondary guide and induces B_4' and B_3'.

On iris I_2, a small part of the energy (10^{-3}) goes through the secondary guide and induces B_4'' and B_3''.

– In branch 3, if waves B_3' and B_3'' have the same phase, then $B_3 \neq 0$.

– But in branch 4, if waves B_4' and B_4'' are in opposition, then $B_4 \neq 0$.

The difference in way is $2l = \lambda_g / 2$. Then, we have a diphase of π.

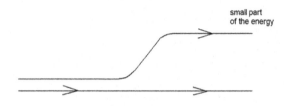

small part
of the energy

Figure 1.5. *Use of coupler*

1.2.4. *Use of couplers*

This system is used to take a previous small part of the microwave energy. It can be used:

– to measure the frequency, power, etc.

– to make a feedback technique and then equalize the output power.

1.3. Lossless equivalent circuit

1.3.1. *Mesh law*

We give the lossless equivalent circuit on a dz length. We know the equivalent circuit in the case of alone line and we suppose two lines are coupled by:

– a magnetic coupling M; and

– a capacitive coupling C_m.

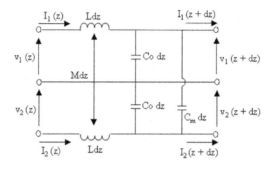

Figure 1.6. *Equivalent circuit of two coupled lines on a dz length*

The mesh law of the equivalent circuit is now:

$$\begin{cases} V_1(z) = Ldz\dfrac{\partial I_1}{\partial t} + Mdz\dfrac{\partial I_2}{\partial t} + V_1(z+dz) \\ V_2(z) = Ldz\dfrac{\partial I_2}{\partial t} + Mdz\dfrac{\partial I_1}{\partial t} + V_2(z+dz) \end{cases}$$

But by definition:

$$\begin{cases} V_1(z) - V_1(z+dz) = -\dfrac{\partial V_1}{\partial z}dz \\ V_2(z) - V_2(z+dz) = -\dfrac{\partial V_2}{\partial z}dz \end{cases}$$

And we get:

$$\begin{cases} -\dfrac{\partial V_1}{\partial z} = L\dfrac{\partial I_1}{\partial t} + M\dfrac{\partial I_2}{\partial t} \\ -\dfrac{\partial V_2}{\partial z} = M\dfrac{\partial I_1}{\partial t} + L\dfrac{\partial I_2}{\partial t} \end{cases}$$

1.3.2. *Notch law*

Now, for the notch law, we have to consider the currents:

$$\begin{cases} I_1(z) = C_0 dz \dfrac{\partial V_1}{\partial t} + C_m dz \dfrac{\partial (V_1 - V_2)}{\partial t} + I_1(z + dz) \\ I_2(z) = C_0 dz \dfrac{\partial V_2}{\partial t} + C_m dz \dfrac{\partial (V_2 - V_1)}{\partial t} + I_2(z + dz) \end{cases}$$

Using:

$$\begin{cases} I_1(z) - I_1(z + dz) = -\dfrac{\partial I_1}{\partial z} dz \\ I_2(z) - I_2(z + dz) = -\dfrac{\partial I_2}{\partial z} dz \end{cases}$$

We also get:

$$\begin{cases} -\dfrac{\partial I_1}{\partial z} = (C_0 + C_m) \dfrac{\partial V_1}{\partial t} - C_m \dfrac{\partial V_2}{\partial t} \\ -\dfrac{\partial I_2}{\partial z} = -C_m \dfrac{\partial V_1}{\partial t} + (C_0 + C_m) \dfrac{\partial V_2}{\partial t} \end{cases}$$

1.3.3. *Coupled differential system of the first order*

Considering the whole capacity:

$$C = C_0 + C_m$$

We have to find a solution of the coupled differential system of four equations:

$$\begin{cases} -\dfrac{\partial I_1}{\partial z} = C \dfrac{\partial V_1}{\partial t} - C_m \dfrac{\partial V_2}{\partial t} & [1.1] \\ -\dfrac{\partial I_2}{\partial z} = -C_m \dfrac{\partial V_1}{\partial t} + C \dfrac{\partial V_2}{\partial t} & [1.2] \\ -\dfrac{\partial V_1}{\partial z} = L \dfrac{\partial I_1}{\partial t} + M \dfrac{\partial I_2}{\partial t} & [1.3] \\ -\dfrac{\partial V_2}{\partial z} = M \dfrac{\partial I_1}{\partial t} + L \dfrac{\partial I_2}{\partial t} & [1.4] \end{cases}$$

1.3.4. *Non-coupled differential system of the second order*

The resolution of these four coupled equations $[1.1]-[1.4]$ is easy.

First, we transform to non-coupled equations. We consider the sums of equations $[1.1]+[1.2]$ and $[1.3]+[1.4]$ and have:

$$\begin{cases} -\dfrac{\partial I_e}{\partial z} = (C - C_m)\dfrac{\partial V_e}{\partial t} \\ -\dfrac{\partial V_e}{\partial z} = (L + M)\dfrac{\partial I_e}{\partial t} \end{cases}$$

With the even (sum) voltages and currents:

$$\begin{cases} I_e = I_1 + I_2 \\ V_e = V_1 + V_2 \end{cases}$$

It is the same as if there is only one line of capacity $(C - C_m)$ and inductance $(L + M)$.

We also consider the differences of equations $[1.1] - [1.2]$ and $[1.3] - [1.4]$ and then we have:

$$\begin{cases} -\dfrac{\partial I_o}{\partial z} = (C + C_m)\dfrac{\partial V_o}{\partial t} \\ -\dfrac{\partial V_o}{\partial z} = (L - M)\dfrac{\partial I_o}{\partial t} \end{cases}$$

With the odd (differences) voltages and currents:

$$\begin{cases} I_o = I_1 - I_2 \\ V_o = V_1 - V_2 \end{cases}$$

It is the same as if there is only one line of capacity $(C + C_m)$ and inductance $(L - M)$.

Combining these two groups of two equations, we obtain equations on the even modes and on the odd modes (non-coupled telegraph equations).

$$\begin{cases} \dfrac{\partial^2}{\partial z^2}\begin{vmatrix} V_e \\ I_e \end{vmatrix} - LC(1+k_L)(1-k_C)\dfrac{\partial^2}{\partial t^2}\begin{vmatrix} V_e \\ I_e \end{vmatrix} = 0 \\[4mm] \dfrac{\partial^2}{\partial z^2}\begin{vmatrix} V_o \\ I_o \end{vmatrix} - LC(1-k_L)(1+k_C)\dfrac{\partial^2}{\partial t^2}\begin{vmatrix} V_o \\ I_o \end{vmatrix} = 0 \end{cases}$$

where we have defined the magnetic and capacitive coefficients and the even and odd speed of propagation:

$$k_L = \frac{M}{L} \text{ and } k_C = \frac{C_m}{C}$$

$$v_e = \frac{1}{\sqrt{LC}\,\sqrt{(1+k_L)(1-k_C)}}$$

$$v_o = \frac{1}{\sqrt{LC}\,\sqrt{(1-k_L)(1+k_C)}}$$

EM state on the two lines results from the superposition of two Transversal Electro Magnetic (TEM) modes. These two modes are orthogonal and they are the normal modes of the coupler.

1.4. Homogeneous medium of permittivity ε

If the medium is homogeneous with a permittivity ε (this is not the case of the microstrip) and the TEM waves are propagating with the speed of the light in this medium, then:

$$v = v_e = v_o = \frac{1}{\sqrt{\mu\varepsilon}}$$

And the coupling coefficient k will be:

$$k = k_L = k_C = \frac{C_m}{C} = \frac{M}{L}$$

We also have:

$$v = \frac{1/\sqrt{LC}}{\sqrt{1-k^2}}$$

where $1/\sqrt{LC}$ is the propagation speed without coupling. Now it is possible to have an expression of the coupling coefficient:

$$k = \sqrt{1 - \frac{\mu\varepsilon}{LC}}$$

1.5. Sinusoidal excitation, even and odd modes

In the case of a sinusoidal excitation $(i = e \text{ and } o)$:

$$\begin{cases} v_i(z,t) = \Re_e\left\{V_i(z)e^{j\omega t}\right\} \\ \mathcal{I}_i(z,t) = \Re_e\left\{I_i(z)e^{j\omega t}\right\} \end{cases}$$

The equations to be satisfied are:

$$\begin{cases} \dfrac{\partial^2}{dz^2}\begin{vmatrix} V_e \\ I_e \end{vmatrix} + \omega^2 LC(1-k^2)\begin{vmatrix} V_e \\ I_e \end{vmatrix} = 0 \\ \dfrac{\partial^2}{dz^2}\begin{vmatrix} V_o \\ I_o \end{vmatrix} + \omega^2 LC(1-k^2)\begin{vmatrix} V_o \\ I_o \end{vmatrix} = 0 \end{cases}$$

The propagation constant is:

$$\beta = \omega\sqrt{LC\left(1-k^2\right)}$$

The general solutions of these two last equations with $(i = e \; and \; o)$ are:

$$\begin{cases} V_i(z) = V_i^+ e^{-j\beta z} + V_i^- e^{j\beta z} \\ I_i(z) = I_i^+ e^{-j\beta z} + I_i^- e^{j\beta z} \end{cases}$$

The first and second parts of these general solutions correspond to progressive (going through z positive) and regressive (going through z negative) waves as shown in Figure 1.7:

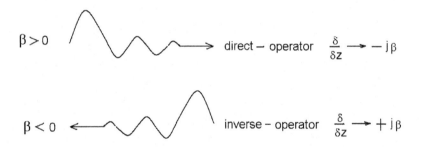

Figure 1.7. *Progressive (direct) and regressive (inverse) waves*

The characteristic impedance of the even mode Z_{oe} is given in the case where $V_o = 0$ and $I_0 = 0$, then:

$$\begin{cases} -\dfrac{\partial I_e}{\partial z} = j\omega(C - C_m)V_e = j\omega C(1-k)V_e \\ -\dfrac{\partial V_e}{\partial z} = j\omega(L + M)I_e = j\omega L(1+k)I_e \end{cases}$$

We consider only the direct wave and get:

$$\begin{cases} j\beta I_e^+ = j\omega C\left(1-k\right)V_e^+ \\ j\beta V_e^+ = j\omega L\left(1+k\right)I_e^+ \end{cases}$$

And we define the characteristic impedances of the even mode Z_{oe} (the impedance the odd modes Z_{oo} will be given using the same method):

$$\begin{cases} Z_{oe} = \dfrac{V_e^+}{I_e^+} = \sqrt{\dfrac{L}{C}\left(\dfrac{1+k}{1-k}\right)} = Z_C\sqrt{\dfrac{1+k}{1-k}} \geq Z_C \\[4mm] Z_{oo} = \dfrac{V_o^+}{I_o^+} = \sqrt{\dfrac{L}{C}\left(\dfrac{1-k}{1+k}\right)} = Z_C\sqrt{\dfrac{1-k}{1+k}} \leq Z_C \end{cases}$$

$Z_C = \sqrt{\dfrac{L}{C}}$ is the characteristic impedance of the uncoupled line. And we have all the time:

$$Z_{oe}.Z_{oo} = Z_C^2$$

1.6. Bibliography

[BAH 03] BAHL I.J., BARTHIA P., *Microwave Solid State Circuit Design*, John Wiley & Sons, May 2003.

[COL 66] COLLIN R.E., *Foundations for Microwave Engineering*, McGraw-Hill, 1966.

[EDW 81] EDWARDS T.C., *Fondations for Microstrip Circuits Design*, John Wiley & Sons, 1981.

[JAR 03] JARRY P., Microwave synthesis of filters and couplers, University of Bordeaux, 2003.

[LEV 66] LEVY R., "Directional couplers", *Advanced in Microwaves*, Academic Press, 1966.

[MAT 70] MATSUMOTO A., *Microwave Filters and Circuits*, Academic Press, New York, 1970.

[PEN 88] PENNOCK S.R., SHEPHERD P.R., *Microwave Engineering with Wireless Applications*, McGraw-Hill Telecommunications, 1988.

2

Strip Coupler

2.1. Introduction

The propagation structure is made up of three conductors in a homogeneous medium (only one dielectric). From Figure 2.1, we can see that two modes can propagate: the symmetrical mode (even mode) and the non-symmetrical mode (odd mode). These are called the normal modes of the coupler.

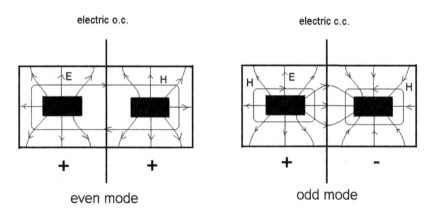

Figure 2.1. *Even and odd modes of the strip coupler*

2.2. Two coupled lines closed on $R_C(Z_C)$

We consider two coupled lines closed on the characteristic impedance R_C as shown in Figure 2.2.

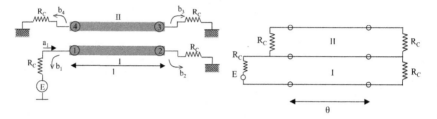

Figure 2.2. *Two coupled lines*

The length of the coupling is $l = \lambda/4$, which corresponds to an electric length of θ:

$$\theta = \beta l = \frac{2\pi}{\lambda} l = \frac{\pi}{2}$$

From Figure 2.3 and using the theorem of the superposition, the result state $(state\, R)$ can be considered as the sum of the even state $(mode\, E)$ and the odd state $(mode\, O)$.

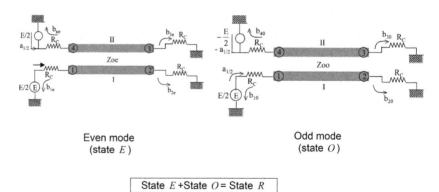

Even mode
(state E)

Odd mode
(state O)

State E +State O = State R

Figure 2.3. *Even and odd modes of two coupled lines*

Now, we come to the study of only one line: the even line (Figure 2.4) or the odd line (Figure 2.5).

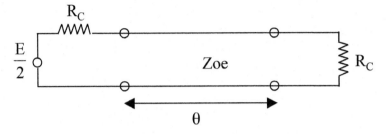

Figure 2.4. *The even line*

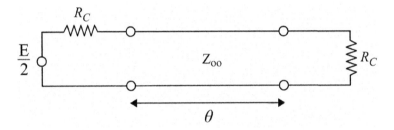

Figure 2.5. *The odd line*

2.3. Reflection (r) and transmission (t) of the even and odd modes

2.3.1. *The reflection (ρ)*

We consider the case of the even mode. The line is characterized by its (A, B, C, D) matrix:

$$\begin{pmatrix} A & B \\ C & D \end{pmatrix} = \begin{pmatrix} \cos\theta & jZ_{oe}\sin\theta \\ \dfrac{j}{Z_{oe}}\sin\theta & \cos\theta \end{pmatrix}$$

This line can be considered as a network of characteristic impedance Z_{oe} closed on impedance Z_L.

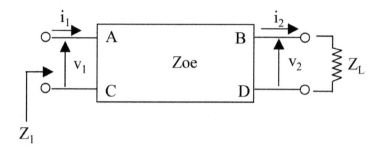

Figure 2.6. *Equivalent network of the even line*

The input impedance and the reflection coefficients are:

$$\frac{Z_1}{Z_L} = \frac{A + B / Z_L}{C\,Z_L + D} \text{ and } \rho = \frac{Z_1 - Z_L}{Z_1 + Z_L}$$

which means:

$$\rho = \frac{A + B/Z_L - C\,Z_L - D}{A + B/Z_L + C\,Z_L + D}$$

2.3.2. The transmission (t)

We do the same for the transmission and we get:

$$t = \frac{2}{A + B/Z_L + CZ_L + D}$$

2.3.3. Reflection and transmission in the even and odd cases

The lengths of the line are $\theta = \beta l$, the loads are $Z_L = R_C$ and the coefficients $(A,\ B,\ C,\ D)$ are:

$$A = \cos\theta; \quad B = j\,Z_{oe}\sin\theta; \quad C = \frac{j}{Z_{oe}}\sin\theta; \quad D = \cos\theta \text{ even case}$$

$$A = \cos\theta; \quad B = j\,Z_{oo}\sin\theta; \quad C = \frac{j}{Z_{oo}}\sin\theta; \quad D = \cos\theta \quad \text{odd case}$$

In the even case by using the normalized characteristic impedance

$$z_{oe} = \frac{Z_{oe}}{R_C}$$

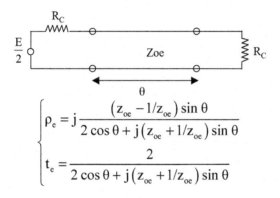

Figure 2.7. *The even line and its reflection and transmission coefficients*

In the odd case by using the normalized characteristic impedance

$$z_{oo} = \frac{Z_{oo}}{R_C}$$

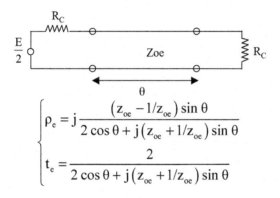

$$\begin{cases} \rho_o = j\,\dfrac{(z_{oo} - 1/z_{oo})\sin\theta}{2\cos\theta + j(z_{oo} + 1/z_{oo})\sin\theta} \\[4mm] t_o = \dfrac{2}{2\cos\theta + j(z_{oo} + 1/z_{oo})\sin\theta} \end{cases}$$

Figure 2.8. *The odd line and its reflection and transmission coefficients*

2.4. Waves of the result state

– State E

In the even case, Figure 2.9 gives the expressions of the even waves $\left(b_{1e}, b_{2e}, b_{3e}, b_{4e}\right)$ as a function of the incident waves of values $+\dfrac{a_1}{2}$ and $+\dfrac{a_1}{2}$.

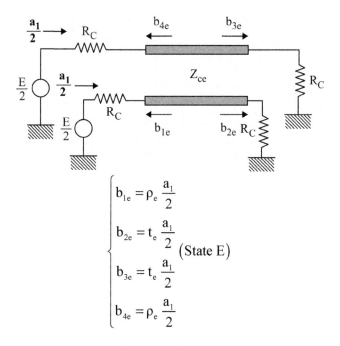

$$\begin{cases} b_{1e} = \rho_e \dfrac{a_1}{2} \\[2mm] b_{2e} = t_e \dfrac{a_1}{2} \\[2mm] b_{3e} = t_e \dfrac{a_1}{2} \\[2mm] b_{4e} = \rho_e \dfrac{a_1}{2} \end{cases} \left(\text{State E}\right)$$

Figure 2.9. *Expressions of the even waves (State E)*

– State O

In the odd case, Figure 2.10 gives the expressions of the odd waves $\left(b_{1o}, b_{2o}, b_{3o}, b_{4o}\right)$ as a function of the incident waves of values $-\dfrac{a_1}{2}$ and $+\dfrac{a_1}{2}$.

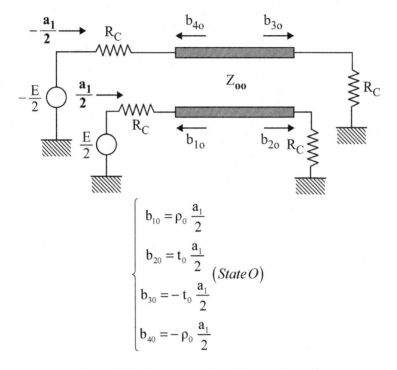

$$\begin{cases} b_{1o} = \rho_0 \dfrac{a_1}{2} \\[2mm] b_{2o} = t_0 \dfrac{a_1}{2} \\[2mm] b_{3o} = -t_0 \dfrac{a_1}{2} \\[2mm] b_{4o} = -\rho_0 \dfrac{a_1}{2} \end{cases} (State\,O)$$

Figure 2.10. *Expressions of the odd waves (State O)*

Result: State R = State E + State O:

The output waves $(b_1,\ b_2,\ b_3,\ b_4)$ are given as a function of the input wave a_1 as follows:

$$\begin{cases} b_1 = b_{1e} + b_{1o} = (\rho_e + \rho_0)\dfrac{a_1}{2} \\[2mm] b_2 = b_{2e} + b_{2o} = (t_e + t_0)\dfrac{a_1}{2} \\[2mm] b_3 = b_{3e} + b_{3o} = (t_e - t_0)\dfrac{a_1}{2} \\[2mm] b_4 = b_{4e} + b_{4o} = (\rho_e - \rho_0)\dfrac{a_1}{2} \end{cases}$$

With $z_{oo} \cdot z_{oe} = 1$ because:

$$z_{oe} = \sqrt{\frac{1+k}{1-k}} \quad ; \quad z_{oo} = \sqrt{\frac{1-k}{1+k}} = \frac{1}{z_{oe}}$$

then:

$$
\begin{cases}
b_1 = 0 \\[2mm]
b_2 = \dfrac{2}{2\cos\theta + j\left(z_{oe} + \dfrac{1}{z_{oe}}\right)\sin\theta}\, a_1 \\[6mm]
b_3 = 0 \\[2mm]
b_4 = j\,\dfrac{\left(z_{oe} - \dfrac{1}{z_{oe}}\right)\sin\theta}{2\cos\theta + j\left(z_{oe} + \dfrac{1}{z_{oe}}\right)\sin\theta}\, a_1
\end{cases}
$$

The constructed device is matched to the generator $(b_1 = 0)$, and there is no energy on port 3 $(b_3 = 0)$. Energy is sent to ports 2 and 4. This device is a directional coupler (Figure 2.11).

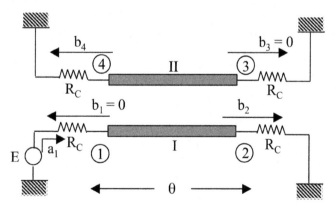

Figure 2.11. *Directional coupler*

Then, a directional coupler gives back energy only on ports 2 and 4:

– port 2 is the direct way;

– port 4 is the coupled way.

2.5. Coupler versus coupling coefficient

We can remark that:

$$z_{oe} - \frac{1}{z_{oe}} = \sqrt{\frac{1+k}{1-k}} - \sqrt{\frac{1-k}{1+k}} = \frac{2k}{\sqrt{1-k^2}}$$

and $z_{oe} + \dfrac{1}{z_{oe}} = \dfrac{2}{\sqrt{1-k^2}}$

$$\begin{cases} b_1 = 0 \\[2mm] b_2 = \dfrac{\sqrt{1-k^2}}{\sqrt{1-k^2}\,\cos\theta + j\sin\theta}\, a_1 \\[2mm] b_3 = 0 \\[2mm] b_4 = j\,\dfrac{k\sin\theta}{\sqrt{1-k^2}\,\cos\theta + j\sin\theta}\, a_1 \end{cases}$$

where k is the coupling coefficient.

2.6. Energy

The energy at the input of the system is:

$$P_1 = \frac{1}{2}|a_1|^2$$

Energy on port 4 is:

$$P_4 = \frac{1}{2}|b_4|^2 = \frac{k^2 \sin^2 \theta}{\left(1-k^2\right)\cos^2 \theta + \sin^2 \theta} \frac{|a_1|^2}{2}$$

$$P_4 = \frac{k^2 \sin^2 \theta}{1-k^2 \cos^2 \theta} \frac{|a_1|^2}{2}$$

Using the same method, energy on port 4 is:

$$P_2 = \frac{1}{2}|b_2|^2 = \frac{1-k^2}{\left(1-k^2\right)\cos^2 \theta + \sin^2 \theta} \frac{|a_1|^2}{2}$$

$$P_2 = \frac{1-k^2}{1-k^2 \cos^2 \theta} \frac{|a_1|^2}{2}$$

We can verify that:

$$P_1 = P_2 + P_4$$

Then, all the energy issued from port 1 goes to ports 2 and 4. And P_2 and P_4 are maximum if $\cos\theta = 0$. This means:

$$\theta = \beta l = \frac{2\pi}{\lambda} l = (2p+1)\frac{\pi}{2}$$

We can give the responses of the directional coupler P_4/P_1 and P_2/P_1 as a function of θ.

We utilize the first band to avoid the superior modes.

Then:

$$\theta = \frac{\pi}{2} \; ; \; l = \frac{\lambda}{4}$$

and:

$$\begin{cases} b_1 = 0 \\ b_2 = -j\sqrt{1-k^2}\,a_1 \\ b_3 = 0 \\ b_4 = k\,a_1 \end{cases}$$

The two waves at the output are in quadrature.

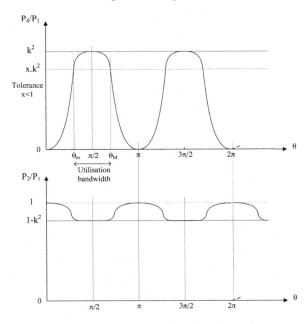

Figure 2.12. *Responses of directional coupler*

Moreover, if we want an equal energy on the two ports 2 and 4:

$$P_2 = P_4$$

which means:

$$1 - k^2 = k^2$$

We must have a coupling coefficient of $k = \dfrac{1}{\sqrt{2}}$ and this is a 3 dB coupler.

Then:

$$\begin{cases} b_1 = 0 \\ b_2 = -\dfrac{j}{\sqrt{2}} a_1 \\ b_3 = 0 \\ b_4 = \dfrac{1}{\sqrt{2}} a_1 \end{cases}$$

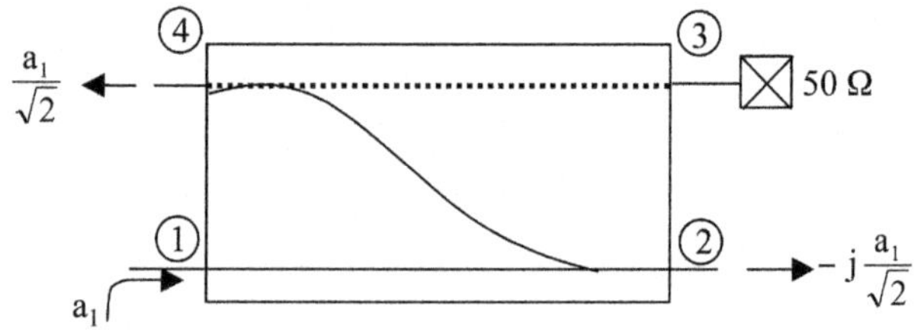

Figure 2.13. *Non-directional coupler with the same energy on ports 2 and 4*

In conclusion, a coupler can be studied by considering only the even mode so that we can determine the transmission coefficient t_e and the reflection coefficient ρ_e (Figure 2.14).

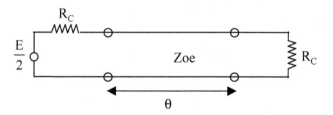

Figure 2.14. *The even mode of the directional coupler*

2.7. Enlargement of the bandwidth

We consider a coupler with n stages with characteristic impedances $Z_{o1}, Z_{o2}, ..., Z_{on}$. This remains to study a line consisted of a cascade of n lines of characteristic impedances $Z_{oe1}, Z_{oe2}, ..., Z_{oen}$. In the case of 3 lines:

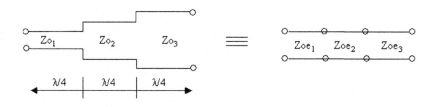

Figure 2.15. *Enlargement of the bandwidth*

2.8. Scattering matrix

A coupler is an octopole with $(a) = (a_1, a_2, a_3, a_4)$ as incident or input waves and $(b) = (b_1, b_2, b_3, b_4)$ as reflected or output waves (Figure 2.16).

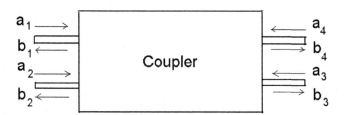

Figure 2.16. *A coupler as an octopole*

And the scattering matrix $(b) = (S)(a)$ will be:

$$
(S) = \begin{pmatrix}
0 & 0 & k & j\sqrt{1-k^2} \\
0 & 0 & j\sqrt{1-k^2} & k \\
k & j\sqrt{1-k^2} & 0 & 0 \\
j\sqrt{1-k^2} & k & 0 & 0
\end{pmatrix}
$$

2.9. Bibliography

[BAH 03] BAHL I.J., BARTHIA P., *Microwave Solid State Circuit Design*, Wiley, 2003.

[CRI 65] CRISTAL E.G., YOUNG L., "Tables of optimum symmetrical TEM-mode coupled transmission-line directional couplers", *IEEE Transactions on Microwave Theory and Techniques*, vol. 13, pp. 544–558, 1965.

[EDW 81] EDWARDS T.C., *Foundations for Microstrip Circuits Design*, John Wiley & Sons, 1981.

[FIR 54] FIRESTONE W.L., "Analysis of transmission line directional couplers", *Proceedings of the IRE*, vol. 42, pp. 1529–1538, October 1954.

[GET 61] GETSINGER W.J., "A coupled strip-line configuration using printed-circuit construction that allows very close coupling", *IRE Transactions on PGMTT*, vol. 9, pp. 535–544, November 1961.

[GET 62] GETSINGER W.J., "Coupled rectangular bars between parallel plates", *IRE Transactions on Microwave Theory and Techniques*, vol. 10, pp. 65–72, 1962.

[JON 56] JONES E.M.T., BOLLJAHN J.T., "Coupled-strip-transmission-line filters and directional couplers", *IRE Transactions on PGMTT*, vol. 4, no. 2, pp. 75–81, April 1956.

[KRA 64] KRAKER D.I., "Asymmetric coupled-transmission-line magic-T", *IEEE Transactions on Microwave Theory and Techniques*, vol. 12, pp. 595–599, 1964.

[LEV 63] LEVY R., "General synthesis of asymmetric multi-element coupled-transmission-line directional couplers", *IEEE Transactions on PGMTT*, vol. 11, pp. 226–237, July 1963.

[LEV 66] LEVY R., "Directional couplers", *Advanced in Microwaves*, Academic Press, 1966.

[LEV 68] LEVY R., LIND L.F., "Synthesis of symmetrical branch-guide directional couplers", *IEEE Transactions on Microwave Theory and Techniques*, vol. 16, pp. 80–89, 1968.

[MAT 70] MATSUMOTO A., *Microwave Filters and Circuits*, Academic Press, New York, 1970.

[OLI 54] OLIVER B.M., "Directional electromagnetic couplers", *Proceedings of the IRE*, vol. 42, pp. 1686–1692, November 1954.

[PEN 88] PENNOCK S.R., SHEPHERD P.R., *Microwave Engineering with Wireless Applications*, McGraw-Hill Telecommunications, 1988.

3

Hybrid and Magic T

3.1. Introduction

Couplers can also be used to characterize the return loss of a structure.

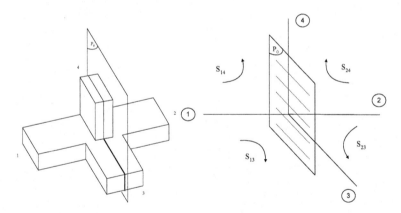

Figure 3.1. *Hybrid T and the symmetries*

There are two methods:

– first, a *zero method* where we compare the unknown reflection coefficient to this of an etalon load;

– second, a *reflectometric method* where we separate and compare the incident and reflected waves.

In this chapter, we will discuss only about the zero methods.

3.2. Coupler hybrid T

A hybrid T is a reciprocal octopol without losses. It has a plane of symmetry (P_o). We use it to characterize the hybrid T by its scattering matrix (S).

First, this element is reciprocal and its scattering matrix is symmetrical:

$$S_{ij} = S_{ji}$$

Then:

$$(S) = \begin{pmatrix} S_{11} & S_{12} & S_{13} & S_{14} \\ S_{12} & S_{22} & S_{23} & S_{24} \\ S_{13} & S_{23} & S_{33} & S_{34} \\ S_{14} & S_{24} & S_{34} & S_{44} \end{pmatrix}$$

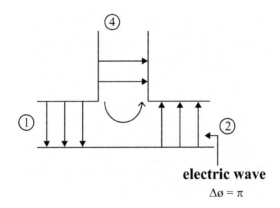

electric wave

$\Delta\emptyset = \pi$

Figure 3.2. *Electric lines at the beginning of the plane (4)*

The symmetry in connection with the plane (P_o) gives:

$$\begin{cases} S_{11} = S_{22} \\ S_{13} = S_{23} \\ S_{14} = -S_{24} \end{cases}$$

Then:

$$(S) = \begin{pmatrix} S_{11} & S_{12} & S_{13} & S_{14} \\ S_{12} & S_{11} & S_{13} & -S_{14} \\ S_{13} & S_{13} & S_{33} & S_{34} \\ S_{14} & -S_{14} & S_{34} & S_{44} \end{pmatrix}$$

Ports 3 and 4 are perfectly uncoupled:

$$S_{34} = 0$$

Then:

$$(S) = \begin{pmatrix} S_{11} & S_{12} & S_{13} & S_{14} \\ S_{12} & S_{11} & S_{13} & -S_{14} \\ S_{13} & S_{13} & S_{33} & 0 \\ S_{14} & -S_{14} & 0 & S_{44} \end{pmatrix}$$

These are the properties of the hybrid T. Now, we give the properties of the magic T.

3.3. Coupler magic T

If the hybrid T is perfectly matched at the four ports:

$$S_{11} = S_{22} = S_{33} = S_{44} = 0$$

We get a magic T characterized by the scattering matrix:

$$(S) = \begin{pmatrix} 0 & S_{12} & S_{13} & S_{14} \\ S_{12} & 0 & S_{13} & -S_{14} \\ S_{13} & S_{13} & 0 & 0 \\ S_{14} & -S_{14} & 0 & 0 \end{pmatrix}$$

We say that this T is also lossless:

$$(S)\left({}^t S^*\right) = (I)$$

$$\begin{pmatrix} 0 & S_{12} & S_{13} & S_{14} \\ S_{12} & 0 & S_{13} & -S_{14} \\ S_{13} & S_{13} & 0 & 0 \\ S_{14} & -S_{14} & 0 & 0 \end{pmatrix} \cdot \begin{pmatrix} 0 & S_{12}^* & S_{13}^* & S_{14}^* \\ S_{12}^* & 0 & S_{13}^* & -S_{14}^* \\ S_{13}^* & S_{13}^* & 0 & 0 \\ S_{14}^* & -S_{14}^* & 0 & 0 \end{pmatrix} = \begin{pmatrix} 1 & 0 & 0 & 0 \\ 0 & 1 & 0 & 0 \\ 0 & 0 & 1 & 0 \\ 0 & 0 & 0 & 1 \end{pmatrix}$$

We get:

$$\begin{cases} |S_{12}| + |S_{13}|^2 + |S_{14}|^2 = 1 \\ |S_{13}|^2 - |S_{14}|^2 = 0 \\ S_{12}^* \, S_{13} = 0 \\ -S_{12}^* \, S_{14} = 0 \end{cases}$$

These relations are possible only if:

$$S_{12} = 0$$

Then:

$$|S_{13}|^2 = |S_{14}|^2$$

and:

$$|S_{13}| = |S_{14}| = \frac{1}{\sqrt{2}}$$

Then:

$$\begin{cases} S_{13} = \dfrac{1}{\sqrt{2}}\, e^{j\varphi} \\[3mm] S_{14} = \dfrac{1}{\sqrt{2}}\, e^{j\phi} \\[3mm] S_{12} = 0 \end{cases}$$

We consider only the real values of S_{13} and S_{14} then their phases are $\varphi = \phi = 0$, and:

$$\begin{cases} S_{13} = S_{14} = \dfrac{1}{\sqrt{2}} \\[3mm] S_{12} = 0 \end{cases}$$

$$(S) = \frac{1}{\sqrt{2}} \begin{pmatrix} 0 & 0 & 1 & 1 \\ 0 & 0 & 1 & -1 \\ 1 & 1 & 0 & 0 \\ 1 & -1 & 0 & 0 \end{pmatrix}$$

The magic T is a perfectly $3\,dB$ directive coupler.

This is a reciprocal, lossless, matched height ports and perfectly decoupled on the arms 1–2 and 3–4.

3.4. Application to the determination of a reflection

3.4.1. *Perfect magic T*

We place on port 4 a match detector $(50\,\Omega)$, and G is the generator.

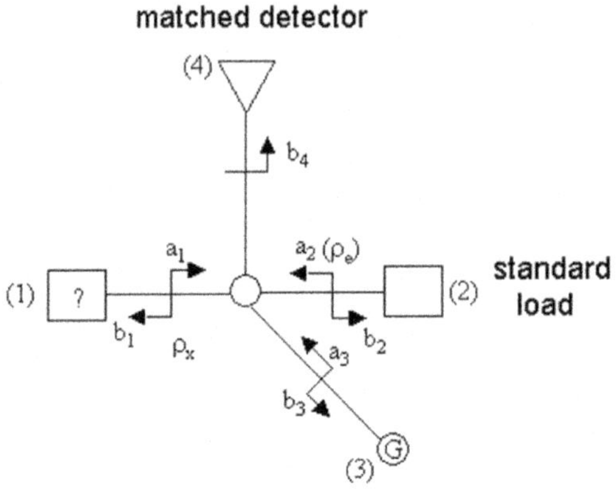

Figure 3.3. *Determination of a reflection with a perfect magic T*

Our problem is to compute b_4 as a function of ρ_e and ρ_x.

From the definition of the scattering matrix of the magic T, we have:

$$
\begin{cases}
b_1 = \dfrac{1}{\sqrt{2}}(a_3 + a_4) \\[2mm]
b_2 = \dfrac{1}{\sqrt{2}}(a_3 - a_4) \\[2mm]
b_3 = \dfrac{1}{\sqrt{2}}(a_1 + a_2) \\[2mm]
b_4 = \dfrac{1}{\sqrt{2}}(a_1 - a_2)
\end{cases}
$$

There is no reflection on the port 4, then $a_4 = 0$.

On the port 1 we have the unknown load, then $a_1 = \rho_x b_1$.

And on the port 2, we place the standard load with $a_2 = \rho_e b_2$.

On the matched detector, we have:

$$b_4 = \frac{1}{\sqrt{2}}(a_1 - a_2)$$

or

$$b_4 = \frac{1}{\sqrt{2}}(\rho_x b_1 - \rho_e b_2)$$

i.e. with $a_4 = 0$ and using the first two equations:

$$b_4 = \frac{1}{2}(\rho_x - \rho_e)a_3$$

Then the wave that goes through the match detector is zero if:

$$b_4 = 0 \rightarrow \rho_x = \rho_e$$

We have the possibility to determine ρ_x if we have a standard load ρ_e.

3.4.2. Non-perfect magic T

In reality, the magic T is not perfectly matched. We have a hybrid T characterized by the scattering matrix:

$$(S) = \begin{pmatrix} S_{11} & S_{12} & S_{13} & S_{14} \\ S_{12} & S_{11} & S_{13} & -S_{14} \\ S_{13} & S_{13} & S_{33} & 0 \\ S_{14} & -S_{14} & 0 & S_{44} \end{pmatrix}$$

Its fluency graph is shown in Figure 3.4.

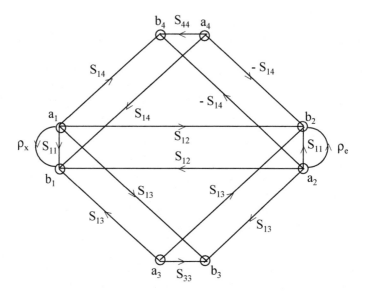

Figure 3.4. *Fluency graph of the non-perfect magic T*

Computing is made using Masson's rule that gives the gain T between the departure notch a_n and the arrival notch b_m.

$$a_n \rightarrow b_m$$

$$T = \frac{b_m}{a_m} = \frac{\sum_{k=1}^{k} T_k \Delta_k}{\Delta}$$

Let us define:

$T_k \Delta_k$ is action chains.

T_k is a transfer functions of the k way.

Δ is a function of the loops.

$\Delta = 1 -$ (sum of all the loops of the network).

+ (sum of the products of the gains of *two* loops *without any contact*).

- (sum of the products of the gains of *tree* loops *without contact*)

+ (......)

$\Delta_k = \Delta -$ (lines or combinations of lines touching the k way).

In this example, the departure is a_3 and the arrival is b_4.

$$\left. \begin{array}{ll} \textit{way } 1 & T_1 = S_{13}\, \rho_e \left(-S_{14}\right) \\ \textit{way } 2 & T_2 = S_{13}\, \rho_x\, S_{14} \\ \textit{way } 3 & T_3 = S_{13}\, \rho_e\, S_{12}\, \rho_x\, S_{14} \\ \textit{way } 4 & T_4 = S_{13}\, \rho_x\, S_{12}\, \rho_e \left(-S_{14}\right) = -T_3 \end{array} \right\} T_k$$

$$\Delta = 1 - (S_{11}\, \rho_x + S_{11}\, \rho_e + S_{12}{}^2\, \rho_e\, \rho_x) + S_{11}{}^2\, \rho_e\, \rho_x$$

$$\begin{cases} \Delta_1 = 1 - S_{11}\, \rho_x & \Delta_3 = 1 \\ \Delta_2 = 1 - S_{11}\, \rho_e & \Delta_4 = 1 \end{cases}$$

then:

$$\frac{b_4}{a_3} = \frac{\Delta_1\, T_1 + \Delta_2\, T_2 + \Delta_3\, T_3 + \Delta_4\, T_4}{\Delta}$$

with:

$$\begin{cases} \varDelta_3 = \varDelta_4 = 1 \\ T_4 = -T_3 \end{cases}$$

then:

$$\frac{b_4}{a_3} = \frac{-\left(1 - S_{11}\,\rho_x\right)S_{13}\,\rho_e\,S_{14} + \left(1 - S_{11}\,\rho_e\right)S_{13}\,\rho_x\,S_{14}}{1 - S_{11}\left(\rho_x + \rho_e\right) + \rho_e\,\rho_x\left(S_{11}^2 - S_{12}^2\right)}$$

And at the final:

$$b_4 = \frac{S_{13}\,S_{14}\left(\rho_x - \rho_e\right)}{1 - S_{11}\left(\rho_x + \rho_e\right) + \rho_e\,\rho_x\left(S_{11}^2 - S_{12}^2\right)}\,a_3$$

to be compared with:

$$b_4 = \frac{1}{2}\left(\rho_x - \rho_e\right)a_3$$

But, we still have:

$$b_4 = 0 \rightarrow \rho_x = \rho_e$$

Using zero detection, a non-perfect magic T gives again the possibility to determine ρ_x if we have a standard load ρ_e.

In this chapter, we have given only the zero methods, it should be interesting to give also the reflectometric methods.

3.5. Bibliography

[BAH 03] BAHL I.J., BARTHIA P., *Microwave Solid State Circuit Design*, John Wiley & Sons, 2003.

[BOU 75] BOUDOURIS G., CHENEVIER P., *Circuits pour Ondes Guidées*, Dunod, 1975.

[EDW 81] EDWARDS T.C., *Foundations for Microstrip Circuits Design*, John Wiley & Sons, 1981.

[HEL 01] HELIER M., *Techniques microondes*, Ellipses, Paris, 2001.

[JAR 03] JARRY P., Microwave Synthesis of Filters and Couplers, University of Bordeaux, 2003.

[LEV 66] LEVY R., "Directional couplers", *Advanced in Microwaves,* Academic Press, New York, 1966.

[MAT 70] MATSUMOTO A., *Microwave Filters and Circuits,* Academic Press, New York, 1970.

[PEN 88] PENNOCK S.R., SHEPHERD P.R., *Microwave Engineering with Wireless Applications,* McGraw-Hill Telecommunications, 1988.

Problems

4.1. Practical determination of the elements of a coupler

We consider a stripline coupler with a dielectric constant $\varepsilon_r = 4$. It works at $f_0 = 6$ GHz with a coupling of $C = -10$ dB and the distance of the two conductors is $b = 3$ mm.

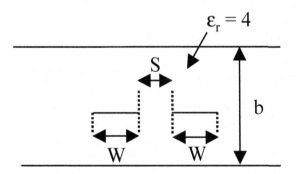

1) What is the value of the coupling coefficient of k?

2) What are the values of the even and odd impedances Z_{oe} and Z_{oo}?

3) What is the value of the coupling length?

The solution

1) We determine k:

$$20\log k = -10$$

$$k = 0.32$$

2) The values of the even and odd impedances are:

$$Z_{oe} = Z_o \sqrt{\frac{1+k}{1-k}} = 69.7 \ \Omega \quad \text{with} \ Z_o = 50 \ \Omega$$

$$Z_{oo} = Z_o \sqrt{\frac{1-k}{1+k}} = 36 \ \Omega$$

Then, using classical curves, we can find the width lines W and their separation S :

From $\sqrt{\varepsilon r} \ Z_{oe} = 139.4 \ \Omega$ and $\sqrt{\varepsilon r} \ Z_{oo} = 72 \ \Omega$, we get $\dfrac{W}{b} = 0.42$

and $\dfrac{S}{b} = 0.06$.

3) If the coupling length is $\lambda_g/4$ that corresponds to $\theta = \pi/2$, then:

$$l = 6.25 \ \text{mm}$$

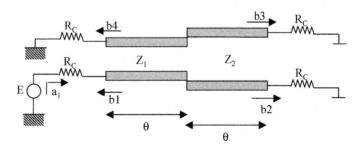

4.2. Two-stages coupler with length $\theta = \pi/2$

A study of this two-stage coupler can be made by considering only the even mode where k_1 and k_2 are the coupling coefficients of the first and second stages.

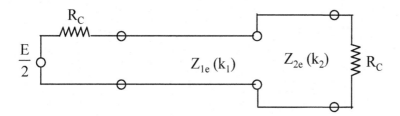

1) What is the matrix (A, B, C, D) of the even mode?

2) What are the transmission and the reflection coefficients of the even modes (t_e, ρ_e) and the odd modes (t_o, ρ_o). We put $\alpha_e = Z_{1e}/Z_{2e}$ and $\alpha_o = Z_{1o}/Z_{2o}$?

3) The normalized even and odd impedances are $z_{ie} \; Z_{ie}/Z_c$ and $z_{io} \; Z_{io}/Z_c$. What is the relation between α_e and α_o?

4) We recall that the waves (b_1, b_2, b_3, b_4) are given by:

$$
\begin{cases}
b_1 = (\rho_e + \rho_0)\dfrac{a_1}{2} \\[2mm]
b_2 = (t_e + t_0)\dfrac{a_1}{2} \\[2mm]
b_3 = (t_e - t_0)\dfrac{a_1}{2} \\[2mm]
b_4 = (\rho_e - \rho_0)\dfrac{a_1}{2}
\end{cases}
$$

What are the values of (b_1, b_2, b_3, b_4) as a function of k_1 and k_2?

5) What is the relation between k_1 and k_2 to have the same energy on ports 2 and 4?

The solution

1) We determine the matrix (A, B, C, D) of the even mode for the first stage:

$$\begin{pmatrix} A & B \\ C & D \end{pmatrix}_1 = \begin{pmatrix} \cos\theta & j\,Z_{1e}\sin\theta \\ j\dfrac{\sin\theta}{Z_{1e}} & \cos\theta \end{pmatrix}$$

if $\theta = \dfrac{\pi}{2}$, then $\begin{pmatrix} A & B \\ C & D \end{pmatrix}_1 = \begin{pmatrix} 0 & j\,Z_{1e} \\ \dfrac{j}{Z_{1e}} & 0 \end{pmatrix}$

Now, for the two stages:

$$\begin{pmatrix} A & B \\ C & D \end{pmatrix} = \begin{pmatrix} A & B \\ C & D \end{pmatrix}_1 \cdot \begin{pmatrix} A & B \\ C & D \end{pmatrix}_2 = \begin{pmatrix} 0 & j\,Z_{1e} \\ \dfrac{j}{Z_{1e}} & 0 \end{pmatrix} \cdot \begin{pmatrix} 0 & j\,Z_{2e} \\ \dfrac{j}{Z_{2e}} & 0 \end{pmatrix}$$

$$\begin{pmatrix} A & B \\ C & D \end{pmatrix} = \begin{pmatrix} -\dfrac{Z_{1e}}{Z_{2e}} & 0 \\ 0 & -\dfrac{Z_{2e}}{Z_{1e}} \end{pmatrix} = \begin{pmatrix} -\alpha_e & 0 \\ 0 & -\dfrac{1}{\alpha_e} \end{pmatrix} \text{ with } \alpha_e = \dfrac{Z_{1e}}{Z_{2e}}$$

We remark that if $Z_{1e} = Z_{2e}$,

then: $\begin{pmatrix} A & B \\ C & D \end{pmatrix} = -\begin{pmatrix} 1 & 0 \\ 0 & 1 \end{pmatrix} = -I$

All is transmitted with a phase of π.

2) (t_e, ρ_e) are calculated using the relations with $B = 0$ and $C = 0$:

$$\begin{cases} t_e = \dfrac{2}{A + B/R_C + CR_C + D} = \dfrac{2}{A + D} \\[3mm] \rho_e = \dfrac{A + B/R_C - CR_C - D}{A + B/R_C + CR_C + D} = \dfrac{A - D}{A + D} \end{cases}$$

Then, with $\alpha_e = \dfrac{Z_{1e}}{Z_{2e}}$, we have:

$$\begin{cases} t_e = \dfrac{-2}{\dfrac{Z_{1e}}{Z_{2e}} + \dfrac{Z_{2e}}{Z_{1e}}} = \dfrac{-2}{\alpha_e + \dfrac{1}{\alpha_e}} \\[6mm] \rho_e = \dfrac{\dfrac{Z_{1e}}{Z_{2e}} - \dfrac{Z_{2e}}{Z_{1e}}}{\dfrac{Z_{1e}}{Z_{2e}} + \dfrac{Z_{2e}}{Z_{1e}}} = \dfrac{\alpha_e - \dfrac{1}{\alpha_e}}{\alpha_e + \dfrac{1}{\alpha_e}} \end{cases}$$

We get the same results in the case of odd mode $\alpha_o = \dfrac{Z_{1o}}{Z_{2o}}$:

$$\begin{cases} t_0 = \dfrac{-2}{\alpha_0 + \dfrac{1}{\alpha_0}} \\[6mm] \rho_0 = \dfrac{\alpha_0 - \dfrac{1}{\alpha_0}}{\alpha_0 + \dfrac{1}{\alpha_0}} \end{cases}$$

3) For the first stage, we have (even and odd modes):

$$z_{1e} = \frac{Z_{1e}}{R_C} = \sqrt{\frac{1+k_1}{1-k_1}} \quad z_{10} = \frac{Z_{10}}{R_C} = \sqrt{\frac{1-k_1}{1+k_1}}$$

The second stage gives (even and odd modes):

$$z_{2e} = \frac{Z_{2e}}{R_C} = \sqrt{\frac{1+k_2}{1-k_2}} \quad z_{20} = \frac{Z_{20}}{R_C} = \sqrt{\frac{1-k_2}{1+k_2}}$$

Then:

$$\alpha_e \cdot \alpha_0 = 1 \text{ and } \alpha_0 = \frac{1}{\alpha_e}$$

4) From the last results, we have:

$$\begin{cases} t_e = \dfrac{-2}{\alpha_e + \alpha_0} \\[3mm] \rho_e = \dfrac{\alpha_e - \alpha_0}{\alpha_e + \alpha_0} \end{cases} \text{and} \begin{cases} t_0 = \dfrac{-2}{\alpha_0 + \alpha_e} = t_e \\[3mm] \rho_0 = \dfrac{\alpha_0 - \alpha_e}{\alpha_0 + \alpha_e} = -\rho_e \end{cases}$$

And we have for the waves (b_1, b_2, b_3, b_4):

$$\begin{cases} b_1 = (\rho_e + \rho_0)\dfrac{a_1}{2} = 0 \text{ matched in } 1 \\[3mm] b_2 = (t_e + t_0)\dfrac{a_1}{2} = t_e\, a_1 = \dfrac{-2}{\alpha_e + \alpha_0}\, a_1 \\[3mm] b_3 = (t_e - t_0)\dfrac{a_1}{2} = 0 \text{ nothing in } 3 \\[3mm] b_4 = (\rho_e - \rho_0)\dfrac{a_1}{2} = \rho_e\, a_1 = \dfrac{\alpha_e - \alpha_0}{\alpha_e + \alpha_0}\, a_1 \end{cases}$$

We have made a coupler with a repartition of the energy to ports 2 and 4 with nothing in 3 and matched in port 1.

Now, we want to find the values of (b_1, b_2, b_3, b_4) as a function of k_1 and k_2.

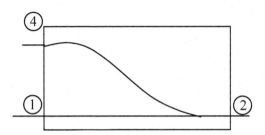

We have:

$$\alpha_e + \alpha_0 = \frac{z_{1e}}{z_{2e}} + \frac{z_{10}}{z_{20}} = \sqrt{\frac{1+k_1}{1-k_1}} \cdot \sqrt{\frac{1-k_2}{1+k_2}} + \sqrt{\frac{1-k_1}{1+k_1}} \cdot \sqrt{\frac{1+k_2}{1-k_2}}$$

Then, we find the expression of $\alpha_e + \alpha_o$ and we can do the same for the difference $\alpha_e - \alpha_o$:

$$\alpha_e + \alpha_0 = 2 \frac{1 - k_1 k_2}{\sqrt{1 - k_1^2} \sqrt{1 - k_2^2}}$$

$$\alpha_e - \alpha_0 = 2 \frac{k_1 - k_2}{\sqrt{1 - k_1^2} \sqrt{1 - k_2^2}}$$

Then:

$$\begin{cases} b_1 \equiv 0 \text{ matched in 1} \\ b_2 = -\dfrac{\sqrt{1 - k_1^2} \sqrt{1 - k_2^2}}{1 - k_1 k_2} a_1 \\ b_3 \equiv 0 \text{ nothing in 3} \\ b_4 = \dfrac{k_1 - k_2}{1 - k_1 k_2} a_1 \end{cases}$$

If $k_2 = 0$, we recover the results with only one stage.

5) Now we impose the same energies P_2 and P_4 issue from ports 2 and $4 P_2 = P_4$

Which means that:

$$\left(1 - k_1^2\right)\left(1 - k_2^2\right) = \left(k_1 - k_2\right)^2$$

Let:

$$k_1 = X$$

We have an equation of the second degree in X as follows:

$$\left(2 - k_2^2\right) X^2 - 2 k_2 X - \left(1 - 2 k_2^2\right) = 0$$

with solutions:

$$X = \frac{k_2 \pm \sqrt{2}\left(1 - k_2^2\right)}{2 - k_2^2}$$

We must have a positive sign because for $k_2 = 0$, we have to recover $X = k_1 = \dfrac{1}{\sqrt{2}}$.

then:

$$k_1 = \frac{k_2 + \sqrt{2}\left(1 - k_2^2\right)}{2 - k_2^2}$$

And the coupler is a 3 dB coupler.

After some manipulations, we get:

$$k_1 - k_2 = \frac{\left(k_2 - \sqrt{2}\right)\left(k_2^2 - 1\right)}{2 - k_2^2}$$

$$1 - k_1 k_2 = \frac{\sqrt{2}\left(1 - k_2^2\right)\left(\sqrt{2} - k_2\right)}{2 - k_2^2}$$

and:

$$\begin{cases} b_1 \equiv 0 \text{ matched in } 1 \\ b_2 = -\dfrac{a_1}{\sqrt{2}} \\ b_3 \equiv 0 \text{ nothing in } 3 \\ b_4 = \dfrac{a_1}{\sqrt{2}} \end{cases}$$

We still have a 3 dB coupler with a larger band.

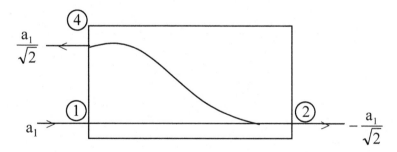

4.3. Perfect directive coupler

We consider a perfect directive coupler with coupling coefficient k and the scattering matrix is given by the relation $(b) = (S)(a)$:

$$(S) = \begin{pmatrix} 0 & 0 & k & j\sqrt{1-k^2} \\ 0 & 0 & j\sqrt{1-k^2} & k \\ k & j\sqrt{1-k^2} & 0 & 0 \\ j\sqrt{1-k^2} & k & 0 & 0 \end{pmatrix}$$

We propose to determine the reflection coefficient ρ_x of an unknown load with comparison to a standard charge ρ_e.

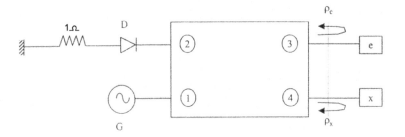

1) The detector is perfectly matched. Give the expression of b_2 as a function of k, ρ_e, ρ_x and a_1 the incident wave of the generator G.

2) What happens to this relation if we have a hybrid coupler $k = \dfrac{1}{\sqrt{2}}$?

3) What is the relation between ρ_x and ρ_e that give a well-balanced bridge $(b_2 = 0)$?

4) If the bridge is not perfectly balanced $(b_2 \neq 0)$, what is the expression of b_1 as a function of k, ρ_e, ρ_x and a_1?

5) What is the expression at the equilibrium?

6) In the case of a hybrid coupler, what is the value of ρ_x that gives a reflection $b_1 = 0$?

The solution

1) We have:

$$\begin{cases} b_2 = j\sqrt{1-k^2}\,a_3 + ka_4 \\ b_3 = ka_1 + j\sqrt{1-k^2}\,a_2 \\ b_4 = j\sqrt{1-k^2}\,a_1 + ka_2 \end{cases}$$

with:

$$\begin{cases} a_3 = \rho_e b_3 \\ a_4 = \rho_x b_4 \end{cases}$$

If the diode is perfect $(a_2 = 0)$, then:

$$\begin{cases} b_3 = ka_1 \\ b_4 = j\sqrt{1-k^2}\,a_1 \end{cases}$$

and:

$$b_2 = j\sqrt{1-k^2}\,\rho_e b_3 + k\rho_x b_4$$

or:

$$b_2 = jk\sqrt{1-k^2}\,\rho_e a_1 + jk\sqrt{1-k^2}\,\rho_x a_1$$

This is:

$$b_2 = jk\sqrt{1-k^2}\,(\rho_e + \rho_x)a_1$$

2) For a hybrid coupler, we have:

$$b_2 = \frac{j}{2}(\rho_e + \rho_x)a_1$$

3) The bridge is well balanced $(b_2 = 0)$ if:

$$\rho_x = -\rho_e$$

4) The matching of the reflectometer:

$$b_1 = ka_3 + j\sqrt{1-k^2}\,a_4$$

$$b_1 = k\rho_e b_3 + j\sqrt{1-k^2}\,\rho_x b_4$$

$$b_1 = k^2 \rho_e a_1 - (1-k^2)\rho_x a_1$$

$$b_1 = [k^2(\rho_e + \rho_x) - \rho_x]a_1$$

5) At the balance $\rho_e = -\rho_x$ and:

$$\frac{b_1}{a_1} = -\rho_x = +\rho_e$$

6) If the coupler is hybrid, $k = \dfrac{1}{\sqrt{2}}$, we have:

$$b_1 = \frac{1}{2}(\rho_e - \rho_x)a_1$$

And the amplitude of b_1 can be zero:

$$b_1 = 0 \quad \text{if} \quad \rho_e = \rho_x$$

The formula is similar for the magic T.

4.4. Bibliography

[BAH 03] BAHL I.J., BARTHIA P., *Microwave Solid State Circuit Design*, John Wiley and Sons, May, 2003.

[EDW 81] EDWARDS T.C., *Fondations for Microstrip Circuits Design*, John Wiley & Sons, 1981.

[HEL 01] HELIER M., *Techniques Microondes*, Ellipses, Paris, 2001.

[JAR 98] JARRY P., KERHERVE E., Lignes Microondes Couplées, Filtres sur Guide, Amplificateurs Microondes, ENSEIRB Bordeaux, 2003.

[LEV 66] LEVY R., "Directional couplers", *Advanced in Microwaves*, Academic Press, 1966.

[MAT 70] MATSUMOTO A., *Microwave Filters and Circuits*, Academic Press, New York, 1970.

[PEN 88] PENNOCK S.R., SHEPHERD P.R., *Microwave Engineering with Wireless Applications*, McGraw-Hill Telecommunications, 1988.

PART 2

Microwave Filters

Analysis of a Guide Resonator with Direct Couplings

5.1. Introduction

We begin to give a very simple "filter" made of only one guide-resonator and two irises. This chapter will give a description and the method of filtering of this system (Figure 5.1).

resonator

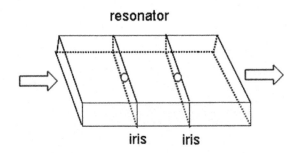

iris   iris

Figure 5.1. *Elementary resonator ("filter")*

The analysis considers:

– the iris alone;

– the resonator (cavity) alone;

– and the resonator between two iris as shown in Figure 5.1.

5.2. Circuit analysis of the iris alone

5.2.1. *Scattering matrix* (S)

This network is lossless and symmetric. From Figure 5.2 with incident unity amplitude (1) there are a transmission (t) and a reflection (ρ) as:

$$1 = t - \rho .$$

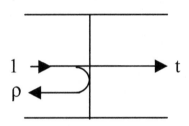

Figure 5.2. *Transmission and reflection on the iris*

The scattering matrix (S) of the iris is symmetrical and lossless, so:

$$(S) = \begin{pmatrix} \rho & t \\ t & \rho \end{pmatrix} \text{ with } (S)^t (S)^* = (I)$$

This gives:

$$\begin{cases} |\rho|^2 + |t|^2 = 1 \\ \rho t^* + \rho^* t = 0 \end{cases}$$

With general solutions:

$$\begin{cases} t = \sin\phi\, e^{-j\psi} \\ \rho = j\cos\phi\, e^{-j\psi} \end{cases}$$

Using the property $t - \rho = 1$ we have:

$$t - \rho = (\sin\phi - j\cos\phi)e^{-j\psi} = -j\,e^{-j(\phi-\psi)} = e^{-j(\phi-\psi-\pi/2)} \equiv 1$$

This is satisfied when:

$$\psi = \phi - \frac{\pi}{2}$$

Then:

$$\begin{cases} \rho = -\cos\phi\, e^{-j\phi} \\ t = j\sin\phi\, e^{-j\phi} \end{cases}$$

ϕ is the opening angle of the diagram. When this angle is closed $\phi = 0$ nothing is transmitted $t = 0$ and all is reflected $\rho = -1$.

and:

$$(S) = e^{-j\phi}\begin{pmatrix} -\cos\phi & j\sin\phi \\ j\sin\phi & -\cos\phi \end{pmatrix}$$

5.2.2. Chain wave matrix (C)

The chain wave matrix is necessary because we have a cascade of elements (iris, guide and iris).

The transformation scattering matrix to chain wave matrix is as follows:

$$(C) = \frac{1}{S_{21}}\begin{pmatrix} 1 & -S_{22} \\ S_{11} & -\Delta S \end{pmatrix}$$

Then we come to:

$$(C) = \frac{1}{t} \begin{pmatrix} 1 & -\rho \\ \rho & t+\rho \end{pmatrix}$$

with ϕ, the opening angle of the diagram:

$$(C) = \frac{j}{\sin\phi} \begin{pmatrix} -e^{j\phi} & -\cos\phi \\ \cos\phi & e^{-j\phi} \end{pmatrix}$$

5.2.3. Tee equivalent circuit

The scattering matrix of a symmetric Tee (Figure 5.3) with impedance ξ is:

$$(S) = \frac{1}{1+2\xi} \begin{pmatrix} -1 & 2\xi \\ 2\xi & -1 \end{pmatrix}$$

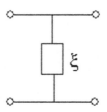

Figure 5.3. *Symmetric Tee*

Comparing with the scattering matrix of the iris, the two matrixes are equivalent if:

$$\xi = j\frac{tg\phi}{2} = \frac{1}{jb}$$

5.3. Circuit analysis of the cavity alone

The cavity is approximated to a lossless line of electric length θ. Its chain matrix is:

$$(C) = \begin{pmatrix} e^{j\theta} & 0 \\ 0 & e^{-j\theta} \end{pmatrix}$$

5.4. Circuit analysis of a cavity between two irises

5.4.1. *Chain wave matrix* (C)

We have a guide of electric length $\theta = \beta l$ where the propagation constant is β and l the length of the cavity.

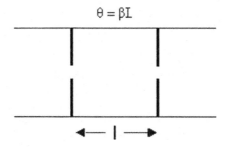

Figure 5.4. *A cavity inserted between two iris*

The two irises are considered to be identical. And we have the equivalent circuit in the Figure 5.5.

The wave chain matrix of the network "iris – cavity – iris" is as following:

$$(C) = \frac{-1}{\sin^2 \varphi} \underbrace{\begin{pmatrix} -e^{j\varphi} & -\cos \varphi \\ \cos \varphi & e^{-j\varphi} \end{pmatrix}}_{\text{1st iris}} \underbrace{\begin{pmatrix} e^{j\theta} & 0 \\ 0 & e^{-j\theta} \end{pmatrix}}_{\text{cavity}} \underbrace{\begin{pmatrix} -e^{j\varphi} & -\cos \varphi \\ \cos \varphi & e^{-j\varphi} \end{pmatrix}}_{\text{2nd iris}}$$

$$(C) = \frac{-1}{\sin^2 \varphi} \begin{pmatrix} e^{j(\theta+2\phi)} - \cos^2 \varphi\, e^{-j\theta} : & 2j\cos\varphi\sin(\theta+\varphi) \\ \cdots\cdots\cdots\cdots\cdots \\ -2j\cos\varphi\sin(\theta+\varphi) & : & e^{-j(\theta+2\phi)} - \cos^2 \varphi\, e^{j\theta} \end{pmatrix}$$

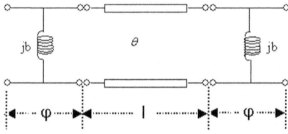

Figure 5.5. *Equivalent circuit of the cavity inserted between two irises*

This can be written as:

$$(C) = \frac{-1}{\sin^2 \varphi} \begin{pmatrix} e^{j\phi}[e^{j(\theta+\phi)} - \cos^2 \varphi\, e^{-j(\theta+\phi)}] : & 2j\cos\varphi\sin(\theta+\varphi) \\ \cdots\cdots\cdots\cdots\cdots \\ -2j\cos\varphi\sin(\theta+\varphi) & : & e^{-j\phi}[e^{-j(\theta+\phi)} - \cos^2 \varphi\, e^{j(\theta+\phi)}] \end{pmatrix}$$

5.4.2. Resonance

The transformation of the chain wave matrix to the scattering matrix is as follows:

$$(S) = \frac{1}{C_{11}} \begin{pmatrix} C_{21} & \Delta C \\ 1 & -C_{12} \end{pmatrix}$$

Then we come to the overall scattering matrix:

$$(S) = \frac{-1}{e^{j\phi}[e^{j(\theta+\phi)} - \cos^2 \varphi\, e^{-j(\theta+\phi)}]} \begin{pmatrix} 2j\cos\varphi\sin(\theta+\varphi) & \sin^2\phi \\ \sin^2\phi & 2j\cos\varphi\sin(\theta+\varphi) \end{pmatrix}$$

At the resonance we need a total reflection zero $S_{11} = 0$ (or $S_{22} = 0$).

$$\cos\phi.\sin(\theta+\phi) = 0$$

Mathematically, the two cases are possible but:

If $\cos\phi = 0$ then $\phi = \dfrac{\pi}{2}$. We eliminate this value of ϕ which corresponds to a solution without iris.

In fact we must have $\sin(\theta+\phi) = 0$. This second solution corresponds at the resonance to:

$$\theta + \phi = k\pi$$

The wave matrix (C) simplifies at the resonance

$$(C) = -\begin{pmatrix} e^{-j\theta} & 0 \\ 0 & e^{j\theta} \end{pmatrix} \text{ and } |C| = \begin{pmatrix} 1 & 0 \\ 0 & 1 \end{pmatrix}$$

All the energy goes through the network "iris – cavity – iris". In general, we use a half wave cavity $(k=1)$.

$$\theta + \phi = \pi$$

5.4.3. The iris at a frequency near the resonance

Near the resonance, we have:

$$\theta = -\phi + \pi + \varepsilon \text{ with } \varepsilon \rightarrow 0$$

But also ϕ is small and we had:

$$\begin{cases} S_{11} \approx \dfrac{2j\sin\varepsilon}{2j\sin\varepsilon - \phi^2} \\[3mm] S_{21} \approx \dfrac{\phi^2}{2j\sin\varepsilon - \phi^2} \end{cases}$$

The electric lengths θ and ϕ depend on the frequency:

$$\sin \varepsilon \cong \varepsilon = \frac{d(\theta+\phi)}{d\omega}\Delta\omega \cong \frac{d\theta}{d\omega}\Delta\omega$$

Because we have:

$$\frac{d\theta}{d\omega} >> \frac{d\phi}{d\omega}$$

If λ_g and l are respectively the length wave and the length of the guide:

$$\sin \varepsilon = \frac{d}{d\omega}\left(\frac{2\pi l}{\lambda_g}\right)\Delta\omega = -\frac{2\pi l}{\lambda_g^2}\frac{d}{d\omega}\left(\lambda_g\right)\Delta\omega$$

but:

$$\frac{1}{\lambda_0^2} = \frac{1}{\lambda_g^2} + \frac{1}{\lambda_c^2} \quad \Rightarrow \quad \frac{d\lambda_g}{\lambda_g} = \left(\frac{\lambda_g}{\lambda_0}\right)^2\frac{d\lambda_0}{\lambda_0}$$

and:

$$\sin \varepsilon = -\frac{2\pi l}{\lambda_g}\cdot\frac{1}{d\omega}\cdot\left(\frac{\lambda_g}{\lambda_0}\right)^2\left(\frac{d\lambda_0}{\lambda_0}\right)\Delta\omega$$

but:

$$\lambda_0 = \frac{C}{f_0} = \frac{2\pi C}{\omega} \text{ is varying with } \omega^{-1}$$

then:

$$\frac{d\lambda_0}{\lambda_0} = -\frac{d\omega}{\omega}$$

and:

$$\sin \varepsilon = \frac{2\pi l}{\lambda_g} \left(\frac{\lambda_g}{\lambda_0}\right)^2 \frac{\Delta\omega}{\omega}$$

But the cavity is half wave $\left(l \approx \dfrac{\lambda_g}{2}\right)$ and we are at a frequency $\omega \approx \omega_0$:

$$\sin \varepsilon = \pi \left(\frac{\lambda_g}{\lambda_0}\right)^2 \frac{\Delta\omega}{\omega_0}$$

We can say that $\sin \varepsilon$ is proportional to the variation $\Delta\omega$ around ω_0.

With the expression of $|S_{11}|$:

$$|S_{11}|^2 = \frac{4\sin^2 \varepsilon}{4\sin^2 \varepsilon - \phi^4}$$

It is possible to compute the VSWR and to see that we have around $\omega = \omega_0$:

$$VSWR(\omega_0 - \Delta\omega) = VSWR(\omega_0 + \Delta\omega)$$

Using:

$$VSWR = \frac{1+|S_{11}|}{1-|S_{11}|}$$

This simple microwave filter made of only two irises and one guide resonator is difficult to study by analysis. That is why, after analysing the irises, we will give a modern method as the synthesis method.

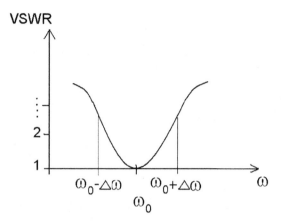

Figure 5.6. *VSWR of the "iris-cavity-iris" around the resonant frequency ω_0*

5.5. Bibliography

[BOU 74] BOUDOURIS G., CHENEVIER P., *Circuits pour Ondes Guidées*, Dunod, 1974.

[COH 57] COHN S.B., "Direct-coupled-resonators filters" *Proc. IRE*, pp. 187–197, February 1957.

[COL 66] COLLIN R.E., *Foundations for Microwave Engineering*, McGraw-Hill, 1966.

[EDW 81] EDWARDS T.C., *Foundations for Microwave Circuits Design*, John Wiley & Sons, 1981.

[FUS 87] FUSCO V.F., *Microwave Circuits – Analysis and Computer-Aided Design*, Printice Hall, 1987.

[JAR 08] JARRY P., BENEAT J., *Advanced Design Techniques and Realizations of Microwave and RF Filters*, Wiley-IEEE Press, 2008.

[MAT 64] MATTHAEI G.L., YOUNG L., JONES E.M.T., *Microwave Filters, Impedance-matching Networks, and Coupling Structures*, McGraw-Hill Book Company, Inc., 1964.

[MAL 79] MALHERBE J.A.G., *Microwave Transmission Line Filters*, Artech House, 1979.

[PEN 98] PENNOCK S.R., SHEPHERD P.R., *Microwave Engineering with Wireless Applications*, McGraw-Hill Telecommunications, 1998.

[RHO 76] RHODES J.D., *Theory of Electrical Filters*, John Wiley & Sons, 1976.

6

Electromagnetic (EM) of the Iris

6.1. Introduction

In the previous chapter, we said that the equivalent circuit of the different iris formed by diaphragms, coupling holes, self bars, etc., is a parallel and imaginary admittance. As an example, we will show that the equivalent circuit of the diaphragm is given by an inductance $j\bar{X}$ (Figure 6.1).

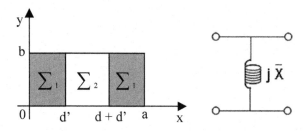

Figure 6.1. *Equivalent circuit of a diaphragm*

6.2. Characterization of the iris

From Figure 6.2, it is clear that the incident mode is the fundamental mode's TE_{10} and the guide is considered as monomode (only the TE_{10} can be propagated).

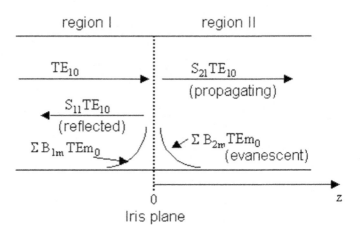

Figure 6.2. *Characterization of the iris*

The transport energy of this incident mode TE_{10} is reflected and transmitted by parts on the two sides of the iris:

– the reflected energy (region I) is proportional to $S_{11}TE_{10}$, where S_{11} is the reflection coefficient and TE_{10} is the inverse mode ("retrograde");

– the other part of the energy (region II) is transmitted $S_{12}TE_{10}$, where S_{12} is the transmitted coefficient and TE_{10} is the direct mode.

The system is lossless and there is energy preservation:

$$|S_{11}|^2 + |S_{12}|^2 = 1$$

But due to the discontinuity of the iris, there is coupling with the modes TE_{m0} $(m \geq 1)$. We show also by symmetry reason that only the modes TE_{m0} $(m \geq 1)$ can propagate. Also because of the geometry of the guide, the modes TE_{m0} ($m \geq 1$) are evanescent and their amplitude decreases very rapidly.

By recapitulation in the two regions, we have:

– region I, before the iris. The fundamental wave TE_{10}, the reflected wave $S_{11}TE_{10}$ and the evanescent modes $\sum B_{1m}TE_{m0}$;

– region II, after the iris. The transmitted wave $S_{21}TE_{10}$ and the evanescent modes $\sum B_{2m}TE_{m0}$.

To study the discontinuity (iris), we have to determine the coefficients:

$$\begin{cases} S_{11} \\ S_{12} \\ B_{1m} \\ B_{2m} \end{cases}$$

In general, two irises are separated by a half wave. Only the two coefficients S_{11} and S_{12} are interesting.

6.3. Properties of the TE_{m0} modes

We have to consider the propagating and the evanescent waves and also the different powers.

6.3.1. Propagating modes $(m=1)$

These modes are characterized by the two waves E_y and H_x.

$$E_{y1} = -\frac{A_1 k}{\beta_{10}} H_{x1}$$

In fact, we have propagative waves and the two waves need to be multiplied by the propagation term where β_{10} is the propagation constant:

$$e^{-j\beta_{10}x}$$

A_1 is a constant to be determined with the power properties. The cutoff is given by the smallest value of:

$$k_{C\,m,n}^2 = k^2 - \beta_{m,n}^2 = \left(\frac{m\pi}{a}\right)^2 + \left(\frac{n\pi}{b}\right)^2$$

which is for $m = 0$ and $n = 1$.

k is the propagation constant of the plane waves and $\beta_{m,n}$ is the propagation constant of the guide waves.

By normalization, we have:

$$e_1 = -\frac{A_1 k}{\beta_{10}} h_1$$

6.3.2. Evanescent modes (m ≥ 2)

These modes are characterized by the two waves E_y and H_x.

$$E_{y\,m} = -j\frac{A_m k}{\gamma_{m0}} H_{x\,m}$$

In fact, these two evanescent waves are multiplied by $e^{-\gamma_{m,0}z}$, where $\gamma_{m,0}$ is the attenuation constant.

By normalizing, we get:

$$e_m = -j \frac{A_m k}{\gamma_{m0}} h_m$$

6.3.3. *Transport power*

TE_{10} is a propagating wave; then, the Poynting vector is:

$$\int_S \underbrace{(\vec{E}_1 \wedge \vec{H}_1^*)}_{POYNTING} . \vec{u}_z ds = \int_S (\vec{E}_{y_1} \wedge \vec{H}_{x_1}) . \vec{u}_z \, ds = \int_S e_1 \, h_1^* \, ds = 1 \qquad [6.1]$$

This quantity is real and we have an active power. The constant A_1 is determined so that the quantity [6.1] is verified.

But the TE_{m0} are evanescent waves; then in this case:

$$\int_S (\vec{E}_m \wedge \vec{H}_m^*) . \vec{u}_z ds = \int_S e_m \, h_m^* \, ds = j \qquad [6.2]$$

This quantity is complex and we have a reactive power. The constant A_m is determined so that the quantity [6.2] is verified.

Orthogonality of the modes gives us:

$$\int_S (\vec{E}_i \wedge \vec{H}_j^*) . \vec{u}_z \, ds = \int_S e_i \, h_j^* \, ds = 0 \ \ \forall i, j \ \ with \ \ i \neq j$$

6.4. Continuities of the waves

We have two continuity equations: one for the electric waves and the other for the magnetic waves.

$$\left(1 + S_{11}\right) e_1 + \sum_2^\infty B_{1m} \, e_m = S_{12} \, e_1 + \sum_2^\infty B_{2m} \, e_m = \hat{E}y \qquad [6.3]$$

$$\left(1 - S_{11}\right) h_1 - \sum_2^\infty B_{1m} \, h_m = S_{12} \, h_1 + \sum_2^\infty B_{2m} \, h_m \qquad [6.4]$$

The minus sign in equation [6.4] is due to the fact that the waves are going toward the negative z.

E_y is the electric wave in the iris plane; it is a common wave to regions (I) and (II).

Iris is a two-port with symmetrical diffraction. Then making the scalar product $\int_S 3 \, h_1^* \, ds$:

$$\left(1+S_{11}\right) \int_S e_1 \, h_1^* \, ds + \sum_2^\infty B_{1m} \int_S e_m \, h_1^* \, ds = S_{12} \int_S e_1 \, h_1^* \, ds + \sum_2^\infty B_{2m}$$

$$\int_S e_m \, h_1^* \, ds = \int_{\Sigma_2} \hat{E}_y \, h_1^* \, ds$$

Because the wave $\hat{E}_y = 0$ on the two surfaces $\sum_1$.

Using the orthogonality relations:

$$\int_S e_1 \, h_1^* \, ds = 1$$

and:

$$\int_S e_m \, h_1^* \, ds = 0 \quad \forall \, m > 1$$

we get:

$$1 + S_{11} = S_{12} = \int_{\Sigma_2} \hat{E}y \, h_1^* \, ds \tag{6.5}$$

In the same manner, we make

$$\int_S 3\, h_m^* \, ds \ :$$

This gives:

$$B_{1m} = B_{2m} = -j \int_{\Sigma_2} \hat{E}y \, h_m^* \, ds \qquad\qquad [6.6]$$

6.5. Computing the susceptance

We have obtained:

$$\begin{cases} S_{12} = 1 + S_{11} \\ B_{2m} = B_{1m} \end{cases}$$

Then, equation [6.4] can be written as:

$$2\, S_{11}\, h_1 = -2 \sum_2^\infty B_{1m}\, hm \qquad\qquad [6.7]$$

Doing also:

$$\int_S 7^* . \hat{E}y \, ds = \int_{\Sigma_2} 7^* \, \hat{E}y \, ds$$

This means:

$$2\, S_{11}^* \int_{\Sigma_2} h_1^* \, \hat{E}y \, ds = -2 \sum_2^\infty B_{1m}^* \underbrace{\int_{\Sigma_2} h_m^* \, \hat{E}y \, ds}_{\frac{B_{1m}}{-j}}$$

$$2\, S_{11}^* \int_{\Sigma_2} h_1^* \, \hat{E}y \, ds = -2j \sum_2^\infty B_{1m}^* \, B_{1m}$$

or:

$$2\,S_{11}^{*} = \frac{-2j\sum\limits_{2}^{\infty}\left|B_{1m}\right|^{2}}{\int\limits_{\Sigma_{2}} h_{1}^{*}\,\hat{E}y\,ds}$$

that is:

$$2\,S_{11} = \frac{2j\sum\limits_{2}^{\infty}\left|B_{1m}\right|^{2}}{\left(\int\limits_{\Sigma_{2}} h_{1}^{*}\,\hat{E}y\,ds\right)^{*}}$$

But we have with equation [6.5]:

$$1 + S_{11} = \int\limits_{\Sigma_{2}} \hat{E}y\,h_{1}^{*}\,ds$$

and:

$$\frac{1}{\xi} = \frac{-2\,S_{11}}{1 + S_{11}} = \frac{-2j\sum\limits_{2}^{\infty}\left|B_{1m}\right|^{2}}{\left|\int\limits_{\Sigma_{2}} h_{1}^{*}\,\hat{E}y\,ds\right|^{2}}$$

with [6.6]:

$$B_{1m} = -j\int\limits_{\Sigma_{2}} \hat{E}y\,h_{m}^{*}\,ds$$

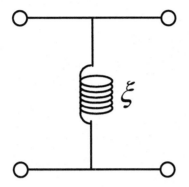

Figure 6.3. *Equivalent circuit of the iris*

The wave's h_1, h_m and $\hat{E}y$ are known, and ξ is also known.

The admittance ξ is then a self as given in Figure 6.3. With this equivalent circuit, we can go back to the filter dimensions.

6.6. Bibliography

[BOU 74] BOUDOURIS G., CHENEVIER P., *Circuits pour Ondes Guidées*, Dunod, 1974.

[COL 60] COLLIN R.E., *Field Theory of Guide Waves*, McGraw-Hill, 1960.

[JAR 03] JARRY P., *Microwave synthesis of filters and couplers*, University of Bordeaux, 2003.

[KON 86] KONG J.A., *Electromagnetic Wave Theory*, John Wiley & Sons, 1986.

[LEW 75] LEWIN L., *Theory of Waveguides*, Butterworth & Co, 1975.

[MAR 51] MARCUVITZ N., *Waveguide Handbook*, McGraw-Hill, 1951.

Synthesis of Guide Filters with Direct Coupling

7.1. What does synthesis mean?

Synthesis is a method of identification between a prototype and a physical structure as shown in the Figure 7.1.

7.2. Description

The filter is formed of n successive cavities. The space between two irises is approximate to $\lambda_{G0} / 2$ as shown in Figure 7.2. We also see in this figure that:

– θ_i is the electric length,

– l_i is the physical length,

– $X_{i,i+1}$ is the normalized reactance of the iris which couples the cavities i and $i+1$.

7.3. The realizations of the iris

The iris are formed by diaphragms, coupling holes, self bars and all have an equivalent circuit given by a parallel and imaginary admittance $j\bar{X}$ (Figure 7.3). As an example using coupling holes we form a filter with direct coupling (Figure 7.3).

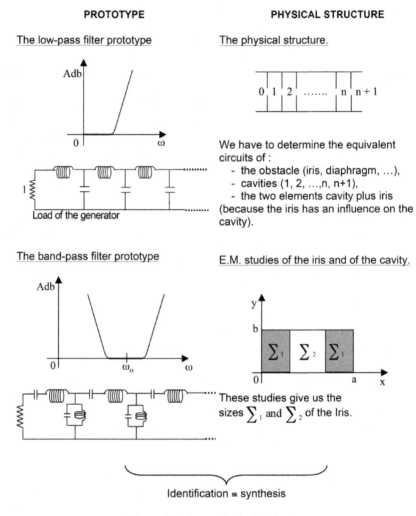

PROTOTYPE

The low-pass filter prototype

PHYSICAL STRUCTURE

The physical structure.

We have to determine the equivalent circuits of :
- the obstacle (iris, diaphragm, ...),
- cavities (1, 2, ...,n, n+1),
- the two elements cavity plus iris (because the iris has an influence on the cavity).

The band-pass filter prototype

E.M. studies of the iris and of the cavity.

These studies give us the sizes $\sum_1$ and $\sum_2$ of the Iris.

Identification ≡ synthesis

Figure 7.1. *The method of synthesis*

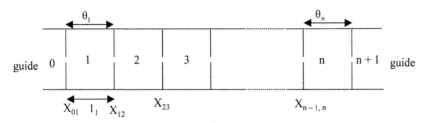

Figure 7.2. *A filter with direct couplings*

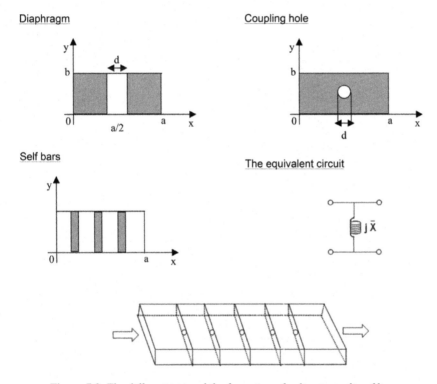

Figure 7.3. *The different iris and the formation of a direct coupling filter*

7.4. Synthesis method of the filter

The synthesis is due to Cohn [COH 57]. It uses a low pass prototype associated with impedance inverters. This method is simple and satisfactory for responses with band ratio less than 15%. An inverter of impedance is a two-port which is characterized by an *ABCD* matrix:

$$\begin{pmatrix} 0 & jK \\ j/K & 0 \end{pmatrix}$$

The input impedance is the inverse of the load impedance:

$$Z_{in} = K^2 / Z_{out}$$

The normalized low pass prototype (Figure 7.4) is now modified to introduce the impedance inverters (Figure 7.5).

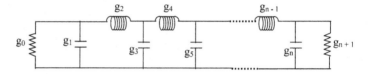

Figure 7.4. *The low-pass prototype*

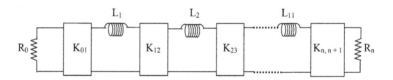

Figure 7.5. *Introduction of impedance inverters* $K_{i,i+1}$

The impedance inverters $K_{i,i+1}$ are a function of the old elements g_i and the new ones L_i :

$$K_{01} = \sqrt{\frac{R_0 L_1}{g_0 g_1}} \; ; \quad K_{i,i+1} = \sqrt{\frac{L_i L_{i+1}}{g_i g_{i+1}}} \; ; \quad K_{n,n+1} = \sqrt{\frac{L_n R_{n+1}}{g_n g_{n+1}}}$$

Choice of elements L_i is then arbitrary. Takahasi's formulas give the value of the elements g_i in the cases of the approximations of Butterworth (maximally flat in the pass-band) and Tchebycheff (equiripple amplitude in the pass-band) [TAK 51].

7.4.1. *Butterworth response*

The Butterworth response is a response which is maximally flat in the pass-band. Then the elements g_i are very simple:

$$
\begin{cases}
g_0 = 1 \\
g_i = 2\sin\dfrac{2i-1}{n} \\
g_n = 1
\end{cases}
$$

7.4.2. *Tchebycheff response*

In the case of a Tchebycheff response, we have ripples in the pass-band and we have an equiripple amplitude in the pass-band. The values of the elements g_i are more complicated but the amplitude is more stringent in the stop-band.

$$
\begin{cases}
g_0 = 1 \\
g_1 = \dfrac{2}{\eta}\sin\dfrac{\pi}{2n} \\
g_i \cdot g_{i+1} = 4\dfrac{\sin\left(2i-1\right)\dfrac{\pi}{2n}.\sin\left(2i+1\right)\dfrac{\pi}{2n}}{\eta^2 + \sin^2\dfrac{i\pi}{n}} \\
g_{n+1} = \left(\sqrt{\varepsilon} + \sqrt{1+\varepsilon}\right)^2 \quad n\, even \\
g_{n+1} = 1 \quad n\, odd
\end{cases}
$$

and $\eta = sh\left(\dfrac{1}{n}sh^{-1}\left(\dfrac{1}{\varepsilon}\right)\right)$

where ε characterizes the ripple in the pass-band (Figure 7.6).

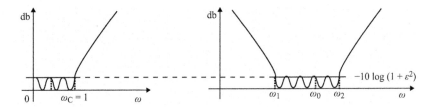

Figure 7.6. *Low-pass and band-pass responses with the ripple*

7.4.3. Pass-band response

The band-pass response is obtained from the frequency transformation:

$$p \rightarrow \alpha \left(\frac{p}{\omega_0} + \frac{\omega_0}{p} \right)$$

where α is the report of the band:

$$\alpha = \frac{\omega_0}{\omega_2 - \omega_1}$$

and ω_0 is the square root of the product of the band edges ω_1 and ω_2 or the half sum of these same band hedges in the case of a narrow band:

$$\omega_0 = \sqrt{\omega_1 \omega_2} \approx \frac{\omega_1 + \omega_2}{2}$$

The network which represents this circuit is given in Figure 7.7.

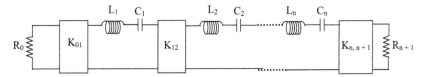

Figure 7.7. *Band-pass prototype*

The impedance inverters $K_{i,i+1}$ will represent the iris while the couples (L_i, C_i) will be the guides. We have now:

$$K_{01} = \sqrt{\frac{\omega_0 R_0 L_1}{\alpha g_0 g_1}} \; ; \quad K_{i,i+1} = \frac{\omega_0}{\alpha} \sqrt{\frac{L_i L_{i+1}}{g_i g_{i+1}}} \; ; \quad K_{n,n+1} = \sqrt{\frac{\omega_0 L_n R_{n+1}}{\alpha g_n g_{n+1}}}$$

with $L_i . C_i = \dfrac{1}{\omega_0^2}$

7.5. Cavity simulation

The localized resonators (L_i, C_i) have to take the place of the distributed ones $X_i(\omega)$ to simulate the microwave cavities.

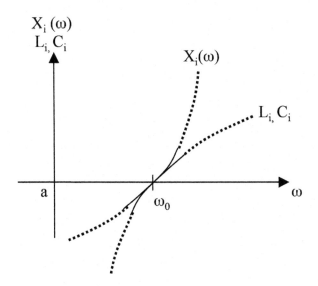

Figure 7.8. *Distributed and localized reactances*

Localized and distributed reactances have to be equal at all the frequencies. It is not possible and identification is made only at the frequency ω_0 (Figure 7.8). The two reactances (localized and distributed) are zero at ω_0, then we force their first derivative to be equal at this resonance ω_0. We have a normalized derivative at ω_0:

$$x_i = \frac{\omega_0}{2} \frac{dX_i(\omega)}{d\omega}\bigg|_{\omega=\omega_0}$$

The normalized reactance of guide of $l = \lambda_g / 2$ length is:

$$X_i(\omega) = tg(\frac{2\pi l_i}{\lambda_g}) = tg(\beta l_i)$$

Then:

$$\frac{dX_i(\omega)}{d\omega} = -\frac{2\pi l_i}{\lambda_g^2 \cos^2(\frac{2\pi l_i}{\lambda_g})} \cdot \frac{d\lambda_g}{d\omega}$$

Using the dispersion:

$$\frac{1}{\lambda^2} = \frac{1}{\lambda_g^2} + \frac{1}{\lambda_c^2} \quad \Rightarrow \quad \frac{d\lambda_g}{\lambda_g} = \left(\frac{\lambda_g}{\lambda}\right)^2 \frac{d\lambda}{\lambda}$$

where λ_g is the length wave in the guide, λ_c is the cut length wave in this guide and λ is the considering length wave which varies in ω^{-1} then:

$$\frac{d\lambda_g}{\lambda_g} = -\left(\frac{\lambda_g}{\lambda}\right)^2 \frac{d\omega}{\omega}$$

The first derivative of the localized and the distributed reactances are equal at the frequency ω_0 (and $\beta l_i = \pi$) if:

$$x_i = L_i \omega_0 = \frac{1}{C_i \omega_0} = \frac{\pi}{2}\left(\frac{\lambda_{g0}}{\lambda_0}\right)^2$$

The network which represents this circuit is given by the Figure 7.9.

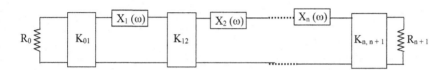

Figure 7.9. *Band-pass prototype with the cavities simulations*

With the new values of the impedance inverters $K_{i,i+1}$ and of the $X_i(\omega)$:

$$K_{01} = \sqrt{\frac{R_0\, x_1}{\alpha\, g_0\, g_1}} \; ; \quad K_{i,i+1} = \frac{1}{\alpha}\sqrt{\frac{x_i\, x_{i+1}}{g_i\, g_{i+1}}} \; ; \quad K_{n,n+1} = \sqrt{\frac{x_n\, R_{n+1}}{\alpha\, g_n\, g_{n+1}}}$$

with $x_i = \dfrac{\pi}{2}\left(\dfrac{\lambda_{g0}}{\lambda_0}\right)^2$

λ_{g0} : guided length wave at ω_0

λ_0 : length wave at ω_0

7.6. Coupling iris simulation

The coupling iris will be represented by the elements $K_{i,i+1}$. These impedance invertors are not realizable with only passive elements. These realizations need purely "negative" elements that will be absorbed by the neighbor elements. Then the $K_{i,i+1}$ are depending on the frequency; that is why the theory is only valuable for a pass-band under 15%.

The coupling iris is purely a self (L) and we choose the next circuit which has to be equivalent to an impedance inverter (K).

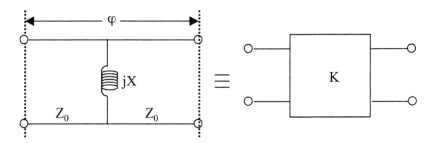

Figure 7.10. *The iris simulations*

We have a reactance (jX) between two lines of characteristic impedances Z_0 and of lengths $\phi/2$. The whole *ABCD* matrix must be:

$$\frac{1}{1+tg^2\left(\dfrac{\phi}{2}\right)}\begin{pmatrix} 1 & jZ_0 tg\left(\dfrac{\phi}{2}\right) \\ \dfrac{j}{Z_0}tg\left(\dfrac{\phi}{2}\right) & 1 \end{pmatrix}\begin{pmatrix} 1 & 0 \\ 1/jX & 1 \end{pmatrix}\begin{pmatrix} 1 & jZ_0 tg\left(\dfrac{\phi}{2}\right) \\ \dfrac{j}{Z_0}tg\left(\dfrac{\phi}{2}\right) & 1 \end{pmatrix}=\begin{pmatrix} 0 & jK \\ 1/jK & 0 \end{pmatrix}$$

This is verified only in the case of a negative length ϕ and at the same time for given values of X as:

$$\phi=-tg^{-1}\left(\frac{2X}{Z_0}\right)$$

$$\left|\frac{X}{Z_0}\right|=\frac{K/Z_0}{1-\left(K/Z_0\right)^2}$$

7.7. Lengths of the cavities

Realization of impedance invertors $X_i(\omega)$ needs "negative" phases which will be compensated by the neighbor elements $\phi_{i-1,i}/2$ and $\phi_{i,i+1}/2$ (Figure 7.11).

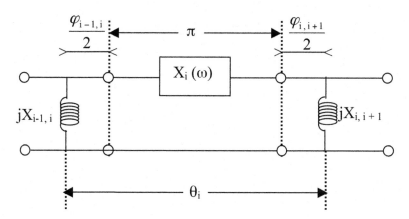

Figure 7.11. *Compensations of the lengths*

θ_i is such that the electric length between the two iris (invertors $jX_{i-1,i}$ and $jX_{i,i+1}$) will be π:

$$\theta_i = \pi + \frac{1}{2}\left[\phi_{i-1,i} + \phi_{i,i+1}\right]$$

This means:

$$\theta_i = \pi - \frac{1}{2}\left[tg^{-1}\left(\frac{2X_{i-1,i}}{Z_0}\right) + tg^{-1}\left(\frac{2X_{i,i+1}}{Z_0}\right)\right]$$

$$i = 1,...,n$$

and the lengths of all the cavities are now known.

7.8. Practical computing of the filter

From a desired amplitude response the values of the elements g_i are given. Then we have to compute successively.

7.8.1. Band ratio

$$\alpha = \frac{\omega_0}{\omega_2 - \omega_1}$$

7.8.2. Invertors

$$\frac{K_{01}}{R_0} = \sqrt{\frac{x}{\alpha R_0 \, g_0 \, g_1}} \; ; \quad \frac{K_{i,i+1}}{R_0} = \frac{x}{\alpha R_0} \frac{1}{\sqrt{g_i \, g_{i+1}}} \; ; \quad \frac{K_{n,n+1}}{R_0} = \sqrt{\frac{x}{\alpha R_0 \, g_n \, g_{n+1}}}$$

with $x = \dfrac{\pi}{2}\left(\dfrac{\lambda_{g0}}{\lambda_0}\right)^2$

λ_{g0} : guided wave length at ω_0

λ_0 : wave length at ω_0

7.8.3. Reactances

$$\frac{X_{i,i+1}}{R_0} = \frac{K_{i,i+1} / R_0}{1 - \left(K_{i,i+1} / R_0\right)^2}$$

$i = 0,...,n$

7.8.4. Lengths of the cavities

$$\theta_i = \pi - \frac{1}{2}\left[tg^{-1}\left(\frac{2 X_{i-1,i}}{R_0}\right) + tg^{-1}\left(\frac{2 X_{i,i+1}}{R_0}\right)\right]$$

$i = 1,...,n$

7.9. Capacitive gap filters

A dual synthesis permits us to design the capacitive gap filters which can be realized in microstrip and at millimetre frequencies.

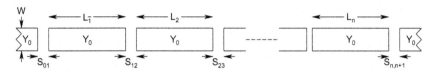

Figure 7.12. *Capacitive gap filter*

7.10. Bibliography

[COH 57] COHN S.B., "Direct-coupled-resonator filters" *Proceedings of IRE*, pp. 187–197, February 1957.

[JAR 03] JARRY P., Microwave synthesis of filters and couplers, University of Bordeaux, 2003.

[JAR 08] JARRY P., BENEAT J., *Advanced Design Techniques and Realizations of Microwave and RF Filters*, Wiley-IEEE Press, 2008.

[MAT 80] MATTHAEI G.L., YOUNG L., JONES E.M.T., *Microwave Filters, Impedance Matching Networks and Coupling Structures*, Artech House, 1980.

[RHO 76] RHODES J.D., *Theory of Electrical Filters*, John Wiley & Sons, 1976.

[TAK 51] TAKAHASI H., "On the ladder type filter network with Tchebycheff response", *Journal of the Institute of Electronics and Communication Engineers of Japan*, vol. 34, pp. 65–74, February 1951.

Problems

8.1. Network formed by identical two-port networks and separated by a guide

We consider a network Q formed by two identical two-port networks q and separated by a lossless diphase guide of length l. This guide is considered as a line of electric length $\theta = \beta_g l$.

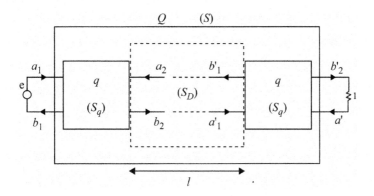

Figure 8.1. *A guide between two identical two-ports*

The networks q are defined by their scattering matrices:

$$\left(S_q \right) = \begin{pmatrix} \rho & t \\ t & \rho \end{pmatrix}$$

We propose to compute the scattering matrix (S) of the whole network Q as a function of the reflection coefficient ρ, the transmission coefficient t and the transmission constant β_g of the guide (line).

1) Let us consider:

$$(S) = \begin{pmatrix} R_1 & T_1 \\ T_2 & R_2 \end{pmatrix}$$

i) What are the relations between R_1, R_2 and T_1, T_2 ?

ii) Give as a function of θ the scattering matrix (S_D) of the diphase guide.

iii) The chain waves matrix (C) is useful to cascade several networks. It gives the waves of the input as a function of the output waves:

$$\begin{pmatrix} b_1 \\ a_1 \end{pmatrix} = \begin{pmatrix} C_{11} & C_{12} \\ C_{21} & C_{22} \end{pmatrix} \begin{pmatrix} a_2 \\ b_2 \end{pmatrix}$$

Compute the chain waves matrix (C_q) of the networks q as a function of ρ and t. What is the chain waves matrix (C_D) of the diphase guide?

iv) What is the reflection coefficient R and the transmission coefficient T of the whole scattering matrix (S) as a function of ρ, t and θ ?

Some comments are given in the particular cases where we have $\theta = 0$, or $\rho = 0$ or $t = 0$.

The solution

i) From Figure 8.1, we have $R_1 = R_2 = R$ and $T_1 = T_2 = T$. Then, the matrix is very simple:

$$(S) = \begin{pmatrix} R & T \\ T & R \end{pmatrix}$$

ii) We consider the diphase guide as a lossless line of electric length $\theta = \beta l$ and:

$$(S_D) = \begin{pmatrix} 0 & e^{-j\theta} \\ e^{-j\theta} & 0 \end{pmatrix}$$

iii) Starting from the definition of the chain waves matrix (C):

$$\begin{pmatrix} b_1 \\ a_1 \end{pmatrix} = \begin{pmatrix} C_{11} & C_{12} \\ C_{21} & C_{22} \end{pmatrix} \begin{pmatrix} a_2 \\ b_2 \end{pmatrix}$$

We must give the input as a function of the output:

$$\begin{cases} b_1 = C_{11} a_2 + C_{12} b_2 \\ a_1 = C_{21} a_2 + C_{22} b_2 \end{cases}$$

(S_q) is a symmetric matrix and we have:

$$\begin{cases} b_1 = \rho a_1 + t a_2 \\ b_2 = t a_1 + \rho a_2 \end{cases}$$

From this second equation, we get:

$$a_1 = -\frac{\rho}{t} a_2 + \frac{1}{t} b_2$$

But the first equation gives:

$$b_1 = -\frac{\rho^2}{t}a_2 + \frac{\rho}{t}b_2 + t\,a_2$$

or

$$b_1 = \frac{t^2 - \rho^2}{t}a_2 + \frac{\rho}{t}b_2$$

and

$$\left(C_q\right) = \frac{1}{t}\begin{pmatrix} t^2 - \rho^2 & \rho \\ -\rho & 1 \end{pmatrix}$$

In the case of the diphase guide, we have:

$$\rho = 0,\, t = e^{-j\theta} \text{ with } \theta = \beta_g l$$

and

$$\left(C_D\right) = \begin{pmatrix} e^{-j\theta} & 0 \\ 0 & e^{j\theta} \end{pmatrix}$$

iv) Then, the overall chain waves matrix is given by:

$$(C) = \left(C_q\right)\left(C_D\right)\left(C_q\right)$$

$$(C) = \frac{1}{t^2}\begin{pmatrix} t^2 - \rho^2 & \rho \\ -\rho & 1 \end{pmatrix}\begin{pmatrix} e^{-j\theta} & 0 \\ 0 & e^{j\theta} \end{pmatrix}\begin{pmatrix} t^2 - \rho^2 & \rho \\ -\rho & 1 \end{pmatrix}$$

$$(C) = \frac{1}{t^2}\begin{pmatrix} (t^2 - \rho^2)^2 e^{-j\theta} - \rho^2 e^{j\theta} & \rho(t^2 - \rho^2)e^{-j\theta} + \rho e^{j\theta} \\ -\rho(t^2 - \rho^2)e^{-j\theta} - \rho e^{j\theta} & -\rho^2 e^{-j\theta} + e^{j\theta} \end{pmatrix}$$

But

$$(C) = \frac{1}{T}\begin{pmatrix} T^2 - R^2 & R \\ -R & 1 \end{pmatrix}$$

Then, we identify:

$$\begin{cases} T = \dfrac{t^2 e^{-j\theta}}{1 - \rho^2 e^{-2j\theta}} \\ R = \rho\dfrac{1 + (t^2 - \rho^2)e^{-2j\theta}}{1 - \rho^2 e^{-2j\theta}} \end{cases}$$

In the particular case of $\theta = 0$:

$$\begin{cases} T = \dfrac{t^2}{1 - \rho^2} \\ R = \rho\dfrac{1 + t^2 - \rho^2}{1 - \rho^2} = \rho(1 + T) \end{cases}$$

If now $\rho = 0$:

$$\begin{cases} T = t^2 e^{-j\theta} \\ R = 0 \end{cases}$$

This means that there is a t^2 transmission with a phase of θ and no reflection.

If $t = 0$:

$$\begin{cases} T = 0 \\ R = \rho \end{cases}$$

There is no transmission and the reflection is ρ.

8.2. Synthesis of a guide filter with direct couplings

We have to realize a direct coupling guide filter with the following specifications.

The center frequency is 3 GHz and the 3 db pass band is 300 MHz. The rejection is 15 db over 600 MHz around the center frequency.

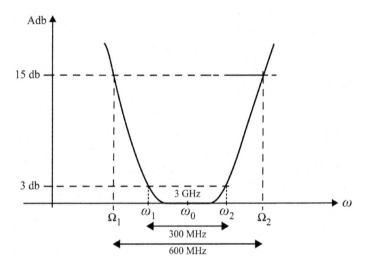

Figure 8.2. *The response of the filter to be realized*

1) Determine the coefficient $1/\alpha$ and find the type of response with the degree n.

2) What are the values of the different normalized inverters $K_{i,i+1}/R_0$?

3) Give the different reactances $X_{i,i+1}$.

4) Compute the length of the cavities L_i.

The solution

1) We have:

$$\frac{1}{\alpha} = \frac{\omega_2 - \omega_1}{\omega_0} = \frac{0.3}{3} = 0.1 = 10\%$$

We are in the case of a (quasi) narrow band. And we return to the low pass (LP) by a translation, and needs a very simple Butterworth filter with $n = 3$.

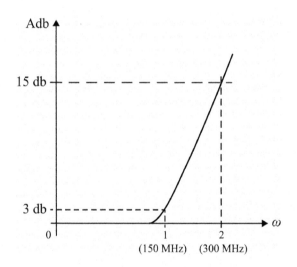

Figure 8.3. *Amplitude of the low-pass prototype*

The solution is maximally flat with an amplitude function:

$$\left| S_{12}(j\omega) \right|^2 = \frac{1}{1 + \omega^6}$$

The synthesis gives the values of the ladder elements g_i.

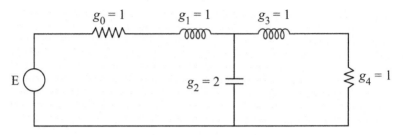

Figure 8.4. *The low-pass prototype with the elements* g_i

Also the factor x is given by:

$$x = \frac{\pi}{2}\left(\frac{\lambda_{g0}}{\lambda_0}\right)^2 \approx \frac{\pi}{2}$$

where the length wave at ω_0 is λ_0 and the guide length wave at the same frequency ω_0 is λ_{g0}.

2) The values of the different normalized inverters $K_{i,i+1}/R_0$ are as follows:

$$\begin{cases} \dfrac{K_{01}}{R_0} = \sqrt{\dfrac{x}{\alpha R_0 g_0 g_1}} \cong \sqrt{\dfrac{\pi}{20}} \cong \sqrt{0.157} \cong 0.4 \\[4mm] \dfrac{K_{12}}{R_0} = \dfrac{x}{\alpha}\dfrac{1}{\sqrt{g_1 g_2}} \cong \dfrac{\pi}{20}\dfrac{1}{\sqrt{2}} \cong 0.157 \\[4mm] \dfrac{K_{23}}{R_0} = \dfrac{x}{\alpha}\dfrac{1}{\sqrt{g_2 g_3}} \cong \dfrac{\pi}{20}\dfrac{1}{\sqrt{2}} \cong 0.157 \\[4mm] \dfrac{K_{34}}{R_0} = \sqrt{\dfrac{x}{\alpha R_0 g_3 g_4}} \cong \sqrt{\dfrac{\pi}{20}} \cong \sqrt{0.157} \cong 0.4 \end{cases}$$

3) The different reactances $X_{i,i+1}$.

$$\begin{cases} \dfrac{X_{01}}{R_0} = \dfrac{K_{01}/R_0}{1-\left(K_{01}/R_0\right)^2} = \dfrac{0.4}{1-\left(0.4\right)^2} = \dfrac{0.4}{0.84} = 0.476 \\[3mm] \dfrac{X_{12}}{R_0} = \dfrac{K_{12}/R_0}{1-\left(K_{12}/R_0\right)^2} = \dfrac{0.157}{1-\left(0.157\right)^2} = \dfrac{0.157}{0.867} = 0.181 \\[3mm] \dfrac{X_{23}}{R_0} = 0.181 \\[3mm] \dfrac{X_{34}}{R_0} = 0.476 \end{cases}$$

4) The different lengths of the cavities L_i are computed using the following formulas:

$$\theta_i = \pi - \frac{1}{2}\left[tg^{-1}\left(\frac{2X_{i-1,i}}{R_0}\right) + tg^{-1}\left(\frac{2X_{i,\,i+1}}{R_0}\right) \right]$$

$i = 1,\dots,4$

and we get:

$$\begin{cases} \dfrac{L_1}{\lambda_{g0}} = 0.412 \\[3mm] \dfrac{L_2}{\lambda_{g0}} = 0.445 \\[3mm] \dfrac{L_3}{\lambda_{g0}} = 0.412 \end{cases}$$

Then, dimensions of the filter guide are discussed in the following.

8.3. Filters using coupled lines: synthesis of S.B. Cohn

This filter structure consists of N resonant lines as shown in Figure 8.6. Each i line of $\lambda_0 / 2$ length is coupled to the $i-1$ line. The length of coupling is $\lambda_0 / 4$. This structure is particularly interesting in the case of integrated and miniaturized filters.

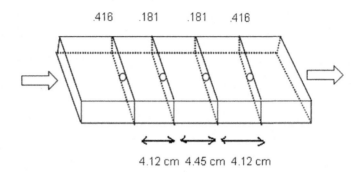

.416   .181   .181   .416

4.12 cm 4.45 cm 4.12 cm

Figure 8.5. *Realization of the filter*

In the following two exercises, we discuss physical properties that can be necessary to learn filters using coupled lines. In this exercise (synthesis of S.B. Cohn), we give a distributed equivalent circuit. The equivalent circuit of two coupled lines can be viewed as a cascaded circuit (Figure 8.7).

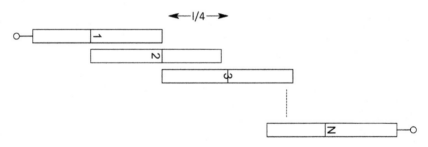

←─ l/4 ─→

Figure 8.6. *Parallel-coupled line structure*

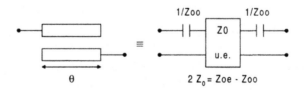

1/Zoo   1/Zoo

Z0

u.e.

θ

$2 Z_0 = Zoe - Zoo$

Figure 8.7. *Equivalent circuit of two coupled lines with a u.e.*

In Figure 8.7, θ is the electric length of the two coupled lines, Z_0 is the characteristic impedance of the lines, Z_{0o} is the impedance of odd mode, Z_{0e} is the impedance of even mode and *u.e.* is the unit element with a line of electric length θ and characteristic impedance Z_0.

1) Give the $(ABCD)$ matrix of a line of electric length θ.

2) What is the value of θ that allows us to identify two coupled lines given by the next circuit?

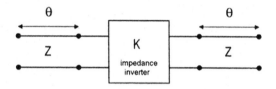

Figure 8.8. *Equivalent circuit with a u.e.*

3) Then, the even and odd impedances (Z_{0e} and Z_{0o}) as a function of Z and K are given. Then, deduce a synthesis method for parallel-coupled line filters.

The solution

1) This is the $(ABCD)$ matrix of a distributed circuit of electric length θ and characteristic impedance Z.

$$\begin{pmatrix} A & B \\ C & D \end{pmatrix} = \begin{pmatrix} \cos\theta & jZ\sin\theta \\ j\dfrac{\sin\theta}{Z} & \cos\theta \end{pmatrix}$$

2) From Figure 8.8, we have a cascaded network and then we use the $(ABCD)$ matrix:

$$
\begin{pmatrix} A & B \\ C & D \end{pmatrix} = \begin{pmatrix} \cos\theta & jZ\sin\theta \\ j\dfrac{\sin\theta}{Z} & \cos\theta \end{pmatrix} \begin{pmatrix} 0 & jK \\ j/k & 0 \end{pmatrix} \begin{pmatrix} \cos\theta & jZ\sin\theta \\ j\dfrac{\sin\theta}{Z} & \cos\theta \end{pmatrix}
$$

Then

$$
\begin{pmatrix} A & B \\ C & D \end{pmatrix} = \begin{pmatrix} -\sin\theta\cos\theta\left(\dfrac{K}{Z}+\dfrac{Z}{K}\right) & jK\cos^2\theta - j\dfrac{Z^2}{K}\sin^2\theta \\ j\dfrac{\cos^2\theta}{K} - j\dfrac{K}{Z^2}\sin^2\theta & -\sin\theta\cos\theta\left(\dfrac{K}{Z}+\dfrac{Z}{K}\right) \end{pmatrix}
$$

The $(ABCD)$ matrix of the equivalent circuit in Figure 8.7 is written as:

$$
\begin{pmatrix} A & B \\ C & D \end{pmatrix} = \begin{pmatrix} 1 & -jZ_{0o}\cot g\theta \\ 0 & 1 \end{pmatrix} \begin{pmatrix} \cos\theta & jZ_0\sin\theta \\ j\dfrac{\sin\theta}{Z_0} & \cos\theta \end{pmatrix} \begin{pmatrix} 1 & -jZ_{0o}\cot g\theta \\ 0 & 1 \end{pmatrix}
$$

or

$$
\begin{pmatrix} A & B \\ C & D \end{pmatrix} = \cos\theta \begin{pmatrix} 1+\dfrac{Z_{0o}}{Z_0} & -jZ_{0o}\left(2+\dfrac{Z_{0o}}{Z_0}\right)\cot g\theta + jZ_0 tg\theta \\ j\dfrac{tg\theta}{Z_0} & 1+\dfrac{Z_{0o}}{Z_0} \end{pmatrix}
$$

3) The two circuits (Figures 8.7 and 8.8) are equivalent if their elements are the same. We have to do the identifications.

For terms A and D and also for terms C and B, we must have:

$$
\begin{cases} -\sin\theta\left(\dfrac{K}{Z}+\dfrac{Z}{K}\right) = 1+\dfrac{Z_{0o}}{Z_0} \\ -K\dfrac{\sin^2\theta}{Z^2}+\dfrac{\cos^2\theta}{K} = \dfrac{\sin\theta}{Z_0} \end{cases}
$$

These two conditions can be satisfied for $\theta = \dfrac{3\pi}{2}$ and we get:

$$\begin{cases} \left(\dfrac{K}{Z} + \dfrac{Z}{K} \right) = 1 + \dfrac{Z_{0o}}{Z_0} \\ Z^2 = K Z_0 \end{cases}$$

Then

$$\begin{cases} Z_{0e} + Z_{0o} = 2\dfrac{Z^2}{K}\left(\dfrac{K}{Z} + \dfrac{Z}{K} \right) \\ Z_{0e} - Z_{0o} = 2\dfrac{Z^2}{K} \end{cases}$$

and we have the well-known formulas:

$$\begin{cases} Z_{0e} = Z\left[1 + \dfrac{Z}{K} + \left(\dfrac{Z}{K} \right)^2 \right] \\ Z_{0o} = Z\left[1 - \dfrac{Z}{K} + \left(\dfrac{Z}{K} \right)^2 \right] \end{cases}$$

Then we give a method of synthesis according to the method of Cohn.

8.4. Filters using coupled lines: synthesis of G.L. Matthaei

In this exercise (synthesis of G.L. Matthaei), we will provide a localized equivalent circuit and after that make an image identification ("identification sur image").

1) Then, we will give the image impedance of the distributed circuit as shown in Figure 8.7.

2) Also the image impedance of the localized circuit is given.

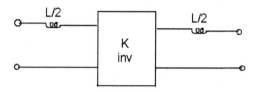

Figure 8.9. *Localized equivalent circuit*

3) Make the identification.

The solution

1) Considering a two-port circuit given by $(ABCD)$ matrix (Figure 8.10).

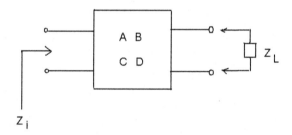

Figure 8.10. *A two-port circuit given by (ABCD) matrix*

The input impedance is:

$$Z_i = \frac{AZ_L + B}{CZ_L + D}$$

and we have the image impedance if $Z_i = Z_L = Z_I$, then:

$$CZ_I^2 - (A - D)Z_I - B = 0$$

The two-port circuit is symmetric when $A = D$ and:

$$Z_I = \sqrt{\frac{B}{C}}$$

We recall that the $(ABCD)$ distributed matrix in Figure 8.7 is written as:

$$\begin{pmatrix} A & B \\ C & D \end{pmatrix} = \cos\theta \begin{pmatrix} 1 + \dfrac{Z_{0o}}{Z_0} & -jZ_{0o}\left(2 + \dfrac{Z_{0o}}{Z_0}\right)\cot g\theta + jZ_0 tg\theta \\ j\dfrac{tg\theta}{Z_0} & 1 + \dfrac{Z_{0o}}{Z_0} \end{pmatrix}$$

This is given as image and distributed impedance:

$$Z_I = \sqrt{\frac{B}{C}} = Z_0 \sqrt{1 - \left[2\left(\frac{Z_{0o}}{Z_0}\right) + \left(\frac{Z_{0o}}{Z_0}\right)^2 \cot g^2\theta\right]}$$

2) The chain matrix of the localized circuit in Figure 8.9 is written as:

$$\begin{pmatrix} A & B \\ C & D \end{pmatrix} = \begin{pmatrix} 1 & j\omega L/2 \\ 0 & 1 \end{pmatrix}\begin{pmatrix} 0 & jK \\ j/K & 0 \end{pmatrix}\begin{pmatrix} 1 & j\omega L/2 \\ 0 & 1 \end{pmatrix}$$

or

$$\begin{pmatrix} -\omega\dfrac{1}{K}\dfrac{L}{2} & jK - j\omega^2\dfrac{1}{K}\left(\dfrac{L}{2}\right)^2 \\ \dfrac{j}{K} & -\omega\dfrac{1}{K}\dfrac{L}{2} \end{pmatrix}$$

and the image impedance of the localized circuit is written as:

$$Z_I = \sqrt{\frac{B}{C}} = \sqrt{K^2 - \left(\frac{\omega L}{2}\right)^2}$$

3) We have 2 degrees of liberty and the identification can be made only at the following two frequencies:

– at the central frequency $\omega = 0$, which corresponds to $\theta = \pi/2$ and gives:

$$Z_0 = K$$

– at the cutoff $\omega = -\omega_1$, which corresponds to $\theta = \theta_1$ and gives:

$$Z_{0o}^2 + 2K Z_{0o} - \left(\frac{\omega_1 L}{2}\right)^2 tg^2\theta_1 = 0$$

By taking the only one positive solution, we get:

$$Z_{0o} = \omega_1 L \left\{ \sqrt{\left(\frac{K}{\omega_1 L}\right)^2 + \left(\frac{tg\theta_1}{2}\right)^2} - \frac{K}{\omega_1 L} \right\}$$

Then, we can perform the synthesis of G.L. Matthaei.

8.5. Bibliography

[CON 58] COHN S.B., "Parallel-coupled transmission line-resonator filters", *IRE Microwave Theory and Techniques*, vol. 6, no. 2, pp. 223–231, April 1958.

[HEL 01] HELIER M., *Techniques Microondes*, Ellipses, Paris, 2001.

[JAR 03] JARRY P., Microwave synthesis of filters and couplers, University of Bordeaux, 2003.

[JAR 08] JARRY P., BENEAT J., *Advanced Design Techniques and Realizations of Microwave and RF Filters*, Wiley-IEEE Press, 2008.

[MAT 60] MATTHAEI G.L., "Design of wide-band (and narrow-band) band-pass microwave filters on the insertion loss basis", *IRE MTT*, vol. 8, no. 6, pp. 580–593, November 1960.

[MAT 64] MATTHAEI G.L., YOUNG L., JONES E.M.T., *Microwave Filters, Impedance-Matching Networks, and Coupling Structures*, McGraw-Hill, 1964.

PART 3

Microwave Amplifiers

Microwave FET Amplifiers and Gains

9.1. Introduction

In many applications of communications, we require amplification of the signals. To avoid obstacles, the communications very often go through a satellite and the conception of reception microwave amplifiers is different from these emission microwave amplifiers. At the emission, the amplitude of the signals is powerful but at reception the amplitude is small (Figure 9.1).

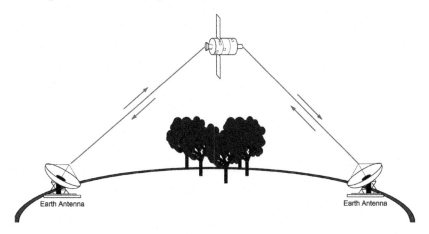

Figure 9.1. *Low and strong power microwave amplifiers*

The characteristics of a microwave amplifier are gain, stability, noise, power, linearity, etc.

We will deal with only the first three properties and give only a measure of the latter two. The choice of the active element will be given by the central frequency f_0, the passband Δf, the gain G and the noise factor F.

In the microwave domain, we use generally field effect transistor (FET) made up of gallium arsenide (GaAs).

From Figure 9.2, we can consider a microwave transistor as a two-port network, which is known by its measured S parameters. The input is the gate, the out is the drain and the source serves as the polarization of the transistor.

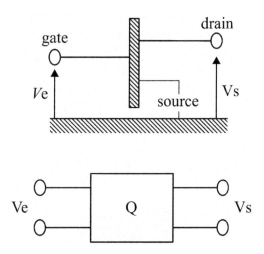

Figure 9.2. *The transistor and its equivalent circuit as a two-port network*

9.2. Recall on the S parameters

At high frequency, we use the S parameters formalism or the elements of the scattering matrix. This is due to the fact that in the microwave domain voltages and currents cannot be directly measured. In general, we use to say that the microwave domain ranges from 1 to 40 GHz .

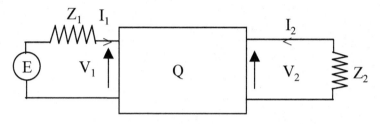

Figure 9.3. *A typical two-port network*

Now we define a_1 and a_2 as the input normalized power waves and b_1 and b_2 as the output normalized power waves.

$$a_1 = \frac{V_1 + Z_1 I_1}{2\sqrt{R_e Z_1}} \qquad b_1 = \frac{V_1 - Z_1^* I_1}{2\sqrt{R_e Z_1}}$$

$$a_2 = \frac{V_2 + Z_2 I_2}{2\sqrt{R_e Z_2}} \qquad b_2 = \frac{V_2 - Z_2^* I_2}{2\sqrt{R_e Z_2}}$$

But in the microwave domain, the impedances Z_1 and Z_2 are equal to 50 Ω:

$$Z_1 = Z_2 = Z_0 = 50\ \Omega$$

and the input and output waves are given by:

$$a_{1,2} = \frac{V_{1,2} + Z_0 I_{1,2}}{2\sqrt{Z_0}}$$

$$b_{1,2} = \frac{V_{1,2} - Z_0 I_{1,2}}{2\sqrt{Z_0}}$$

9.2.1. *The network is sourced and loaded by impedances different from Z_0*

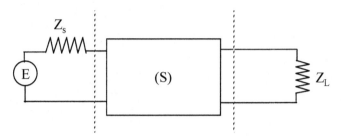

Figure 9.4. *Network with Z_s and Z_L different from Z_0*

In this case, we define the source and load reflections:

$$\rho_s = \frac{Z_s - Z_0}{Z_s + Z_0} \quad \text{and} \quad \rho_L = \frac{Z_L - Z_0}{Z_L + Z_0}$$

9.2.2. *Flow graph of the load*

For example, starting from the first notch a we go to the second notch b through the branch ρ_L.

$$b = a\,\rho_L$$

9.2.3. *Flow graph of the two-port network*

$$b_1 = S_{11}\,a_1 + S_{12}\,a_2$$

$$b_2 = S_{21}\,a_1 + S_{22}\,a_2$$

9.2.4. *Cascade of two two-port networks*

We have at the intersection:

$a_2 = b'_1 \times 1 = b'_1$

$b_2 = a'_1 \times 1 = a'_1$

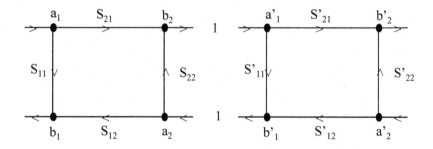

9.2.5. *A load of a two-port network*

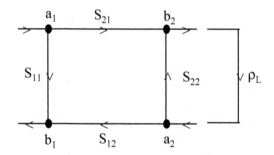

The two-port network is loaded by ρ_L, then:

$a_2 = b_2 \, \rho_L$

9.2.6. *A source*

We define the source notch (b_S), and if this generator is loaded by ρ_L we have:

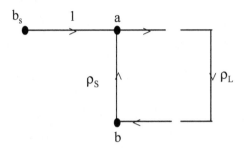

$$\begin{cases} a = b_S \cdot 1 + b\,\rho_S \\ b = a\rho_L \end{cases}$$

Then

$$a = b_s + a\,\rho_L\,\rho_S$$

and

$$a = \frac{b_s}{1 - \rho_L\,\rho_g}$$

9.3. Masson's rules for non-touching loops

Masson says that the gain will be of the form:

$$T = \frac{\sum Tk\,\Delta k}{\Delta}$$

or

$$T = \frac{T_1\left[1 - \sum L^{(1)}_{(1)} + \sum L^{(1)}(2)...\right] + T_2\left[1 - \sum L^{(2)}(1) + \sum L^{(2)}(2)...\right]}{1 - \sum L(1) + \sum L(2)...}$$

where:

$-\sum L(1)$ is the sum of all the first-order loops (way closed on itself described without passing more than one time at every notch);

$-\sum L(2)$ is the sum of all the second-order loops (product of two loops of the first order without touching themselves);

$- T_k$ is the k loops gain;

$-\sum L^{(1)}(1)$ is the sum of the first-order loop without touching T_1 ;

$-\sum L^{(m)}(n)$ is the sum of the nth-order loop without touching T_m ;

$-$ T is the ratio of a dependent variable to an independent variable.

9.4. Transducer power gain of a network with a load and source

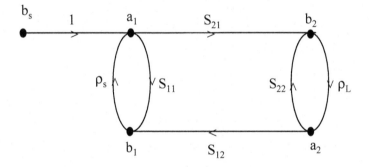

Figure 9.5. *Flow graph of a two-port network loaded by ρ_S, ρ_L and attacked by a source b_S*

This two-port network is characterized by the reflection coefficient of the load ρ_L and the reflection coefficient of the source ρ_S. This two-port network is attacked by a source b_S .

The trajectory goes from b_S to b_2 and the amplitude of the way is $1.S_{21} = S_{21}$.

All the first-order loops are $\rho_S S_{11}$, $\rho_L S_{22}$, $S_{21}\rho_L\ S_{12}\rho_S$, and there is only one second-order loop $\rho_S S_{11}\rho_L S_{22}$.

Then, we have the ratio:

$$T = \frac{b_2}{b_s} = \frac{S_{21}}{1 - [\rho_s\ S_{11} + \rho_L\ S_{22} + S_{12}\ S_{21}\ \rho_L\ \rho_S] + [\rho_S\ \rho_L\ S_{11}\ S_{22}]}$$

or

$$T = \frac{S_{21}}{(1 - S_{11}\ \rho_S)(1 - S_{22}\ \rho_L) - S_{12}\ S_{21}\ \rho_L\ \rho_S}$$

The transducer power gain is written as:

$$G_T = \frac{\text{maximum power delivred to the load}}{\text{available power from source}} = \frac{1}{2}\frac{|b_2|^2\left(1-|\rho_L|^2\right)}{\dfrac{1}{2}\dfrac{|b_S|^2}{\left(1-|\rho_S|^2\right)}}$$

$$G_T = \frac{\left(1-|\rho_L|^2\right)|S_{21}|^2\left(1-|\rho_S|^2\right)}{\left|(1-S_{11}\ \rho_S)(1-S_{22}\rho_L) - S_{12}\ S_{21}\ \rho_L\ \rho_S\right|^2}$$

If the output is matched $\rho_L = 0$, then we have:

$$G_S = \frac{1-|\rho_S|^2}{\left|1-S_{11}\ \rho_S\right|^2}$$

If now the input is matched $\rho_S = 0$, then we have:

$$G_L = \frac{1-|\rho_L|^2}{\left|1-S_{22}\ \rho_L\right|^2}$$

If the two (input and output) are matched $\rho_L = \rho_S = 0$, then we have a simple formula:

$$G_0 = |S_{21}|^2$$

where G_0 is the intrinsic gain.

9.5. Unilateral transducer gain

Unilateral transducer gain (G_{Tu}) is given when the coefficient S_{12} is zero, then we have a perfect symmetric formula. It is the case of the FET amplifiers.

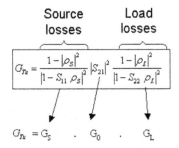

The equation can be written as:

$$G_{Tu} = G_S . G_0 . G_L$$

The G_{Tu} is the perfect product of the losses in the source G_S, of the losses in the load G_L and the intrinsic gain G_0.

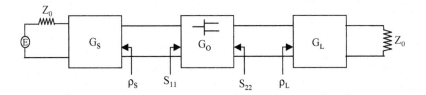

Figure 9.6. *Unilateral amplifier*

G_{Tu} is maximum when there is perfect matching between:

– the input of matching load and the output of the transistor $\rho_L = S_{22}^*$;

– the output of matching source and the input of the transistor $\rho_S = S_{11}^*$.

We have a maximum gain as shown in Figure 9.7.

intrinsic gain

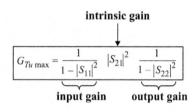

input gain   **output gain**

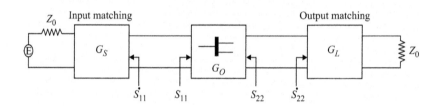

Figure 9.7. *Maximum gain unilateral amplifier*

The $G_{Tu\,max}$ depends only on the S_{ij} of the transistor alone. These values of the $G_{Tu\,max}$ are given by the constructor for different values of the frequency.

9.6. Circles with constant gain (unilateral case $S_{12} = 0$)

In the case of the source, we have a gain G_S and a maximum $G_{S\,max}$:

$$G_S = \frac{1-|\rho_S|^2}{|1-S_{11}\,\rho_S|^2} \quad \text{and} \quad G_{S\,max} = \frac{1}{1-|S_{11}|^2}$$

and a relative source gain is given by:

$$g_s = \frac{G_S}{G_{S\,max}} = G_S \cdot \left(1-|S_{11}|^2\right) = \frac{\left(1-|\rho_S|^2\right)\left(1-|S_{11}|^2\right)}{|1-S_{11}\,\rho_S|^2}$$

but:

$$|1-S_{11}\,\rho_S|^2 = \left(1-S_{11}\,\rho_S\right)\left(1-S_{11}^*\,\rho_S^*\right) = 1+|S_{11}|^2\,|\rho_S|^2 - 2\,R_e\{S_{11}\,\rho_S\}$$

and we get circles with constant gains:

$$|\rho_S|^2 - \frac{2\,g_s}{|S_{11}|^2\,(g_S-1)+1}\,R_e\{S_{11}\,\rho_S\} + \frac{g_s+|S_{11}|^2-1}{|S_{11}|^2\,(g_s-1)+1} = 0$$

is of the form

$$|\rho_S - \Omega_s|^2 = r_S^2$$

It is the parametric equation of circles with center and radius:

$$\Omega_S = \frac{S_{11}^*\,g_S}{1-|S_{11}|^2\,(1-g_s)} \quad \text{Center}$$

$$r_s = \frac{\left(1-|S_{11}|^2\right)\sqrt{1-g_s}}{1-|S_{11}|^2\,(1-g_s)} \quad \text{Radius}$$

In the case of the output, we get the same results when we make the changes:

$$\begin{cases} 1 \to 2 \\ G_S \to G_L \end{cases}$$

In the general case (non-unilateral: $S_{12} \neq 0$), we also get circles but with a different center Ω'_S and different radius r'_S.

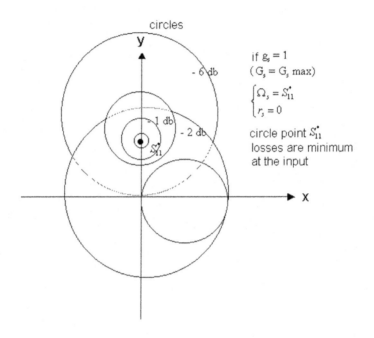

Figure 9.8. *Input constant gain circles*

9.7. Bibliography

[BAH 03] BAHL I.J., BARTHIA P., *Microwave Solid State Circuit Design*, Wiley, 2003.

[BAH 09] BAHL I.J., *Fundamental of RF and Microwave Transistors Amplifiers*, Wiley-Interscience, 2009.

[CHA 94] CHANG K., *Microwave Solid-State Circuits and Applications*, Wiley-Interscience, 1994.

[EDW 81] EDWARDS T.C., *Foundations for Microstrip Circuits Design*, John Wiley & Sons, 1981.

[GEN 84] GENTILI C., *Amplificateurs et Oscillateurs Microondes*, Masson, 1984.

[HA 81] HA T.T., *Solid-State Microwave Amplifier Design*, Wiley-Interscience, 1981.

[JAR 98] JARRY P., KERHERVE E., Lignes Microondes Couplées, Filtres sur Guide, Amplificateurs Microondes, ENSEIRB Bordeaux, 2003.

[JAR 04] JARRY P., Circuits Actifs Microondes: Amplificateurs, Oscillateurs, University of Bordeaux, 2004.

[MAS 53] MASSON S.J., "Feedback theory – some properties of signal flow graphs", *Proceedings of IRE*, vol. 41, pp. 1144–1156, 1953.

[PEN 88] PENNOCK S.R., SHEPHERD P.R., *Microwave Engineering with Wireless Applications*, McGraw-Hill Telecommunications, 1988.

[SOR 88] SOARES R., *GaAs MESFET Circuit Design*, Artech House, 1988.

Stability

10.1. Introduction

Stability depends not only on the amplifier but also on the loads ρ_S and ρ_L. With the values of the scattering parameters S_{ij}, we distinguish the unconditional stability and the conditional stability.

10.2. Unconditional and conditional stabilities

We are in the case of an unconditional stability if there are no conditions on the stability and if the magnitudes of the reflection coefficients of the source S_1 and load S_2 are always less than unity.

$$\forall\, Z_L \left(\text{with } R_e \left\{ Z_L \right\} > 0 \right); \left| S_1 \right| < 1$$

and

$$\forall\, Z_S \left(\text{with } R_e \left\{ Z_S \right\} > 0 \right); \left| S_2 \right| < 1$$

where S_1 is the reflection coefficient side of the source and S_2 is the reflection coefficient side of the load.

In the same way, the network is conditionally stable if there exists some values of Z_1 or Z_2 so that S_1 or S_2 is greater than unity.

$$\forall Z_L \left(\text{with } R_e \{ Z_L \} > 0 \right); \left| S_1 \right| \geq 1$$

or

$$\forall Z_S \left(\text{with } R_e \{ Z_S \} > 0 \right); \left| S_2 \right| \geq 1$$

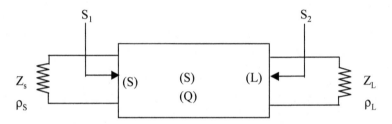

Figure 10.1. *Reflection side input and side output*

But we have:

$$S_1 = S_{11} + \frac{S_{12} \, S_{21} \, \rho_L}{1 - S_{22} \, \rho_L} = \frac{S_{11} \left(1 - S_{22} \, \rho_L \right) + S_{12} \, S_{21} \, \rho_L}{1 - S_{22} \, \rho_L}$$

which gives:

$$S_1 = \frac{S_{11} - \Delta \, \rho_L}{1 - S_{22} \, \rho_L} = \frac{\Delta}{S_{22}} + \frac{B}{1 - S_{22} \, \rho_L}$$

with

$$\Delta = S_{11} \, S_{22} - S_{12} \, S_{21} \text{ and } B_{22} = \frac{S_{12} \cdot S_{21}}{S_{22}}$$

After computing, we have the unconditional stability if:

$$\begin{cases} K = \dfrac{1+|\varDelta|^2 - |S_{11}|^2 - |S_{22}|^2}{2\,|S_{12}\,S_{21}|} > 1 \; (\text{ROLLET}) \\ 1 - |S_{22}|^2 - |S_{12}\,S_{21}| > 0 \; (S_1 \text{ condition}) \\ \text{doing } 1 \leftrightarrow 2 \\ 1 - |S_{11}|^2 - |S_{12}\,S_{21}| > 0 \; (S_2 \text{ condition}) \end{cases}$$

The first is the ROLLET factor and the latter two are the unconditional stability of S_1 and S_2.

These three conditions depend only on the characteristics of the transistor (S_{ij} of the transistor) and they give the unconditional stability for any sort of loads.

10.3. Limits of stability

If these three conditions are not satisfied, the amplificatory is potentially unstable. It is interesting to know the limits of stability.

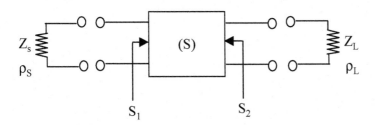

Figure 10.2. *Limits of stability*

We have at the input:

$$S_1 = S_{11} + \frac{S_{12}\,S_{21}\,\rho_L}{1 - S_{22}\,\rho_L} = \frac{S_{11} - \varDelta\,\rho_L}{1 - S_{22}\,\rho_L}$$

The limit of stability is given for $|S_1| = 1$:

$$|S_{11} - \Delta \rho_L|^2 = |1 - S_{22} \rho_L|^2$$

but

$$|S_{11} - \Delta \rho_L|^2 = (S_{11} - \Delta \rho_L)(S_{11}^* - \Delta^* \rho_L^*)$$

and

$$|1 - S_{22} \rho_L|^2 = (1 - S_{22} \rho_L)(1 - S_{22}^* \rho_L^*)$$

which gives:

$$|S_{11}|^2 - 2R_e\left\{S_{11}^* \Delta\rho_L\right\} + |\Delta|^2 |\rho_L|^2 = 1 - 2R_e\left\{S_{22}\rho_L\right\} + |S_{22}|^2 |\rho_L|^2$$

i.e.

$$\left(|S_{22}|^2 - |\Delta|^2\right)|\rho_L|^2 - 2R_e\left\{(S_{22} - \Delta S_{11}^*)\rho_L\right\} + 1 - |S_{11}|^2 = 0$$

The place of ρ_L is a circle of radius r_2 and center Ω_2 so that:

$$|\rho_L - \Omega_2|^2 = |r_2|^2$$

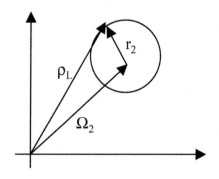

Figure 10.3. *Place of ρ_L*

The equation of the circle is written as:

$$\left(\rho_L - \Omega_2\right)\left(\rho_L^* - \Omega_2^*\right) = \left|r_2\right|^2$$

We get:

$$\left|\rho_L\right|^2 - 2 R_2 \left\{\Omega_2^* \, \rho_L\right\} + \left|\Omega_2\right|^2 = \left|r_2\right|^2$$

To be compared with:

$$\left|\rho_L\right|^2 - 2 R_e \left\{\frac{S_{22} - \Delta\, S_{11}^*}{\left|S_{22}\right|^2 - \left|\Delta\right|^2} \, \rho_L\right\} + \frac{1 - \left|S_{11}\right|^2}{\left|S_{22}\right|^2 - \left|\Delta\right|^2} = 0$$

and with this method we get by identification the center:

$$\Omega_2 = \frac{\left(S_{22} - \Delta\, S_{11}^*\right)^*}{\left|S_{22}\right|^2 - \left|\Delta\right|^2}$$

and the radius is given by:

$$\left|r_2\right|^2 = \frac{\left|S_{22} - \Delta\, S_{11}^*\right|^2}{\left(\left|S_{22}\right|^2 - \left|\Delta\right|^2\right)^2} - \frac{1 - \left|S_{11}\right|^2}{\left|S_{22}\right|^2 - \left|\Delta\right|^2}$$

which is:

$$\left|r_2\right|^2 = \frac{S_{12}\, S_{21} \cdot S_{12}^*\, S_{21}^*}{\left(\left|S_{22}\right|^2 - \left|\Delta\right|^2\right)^2}$$

or

$$r_2 = \frac{\left|S_{12}\right| \cdot \left|S_{21}\right|}{\left|S_{22}\right|^2 - \left|\Delta\right|^2}$$

10.4. Places of stability

The place of the stability is shown in Figure 10.4.

At the center of the SMITH chart, we have:

$$\rho_L = 0 \ or \ |S_1| = |S_{11}|$$

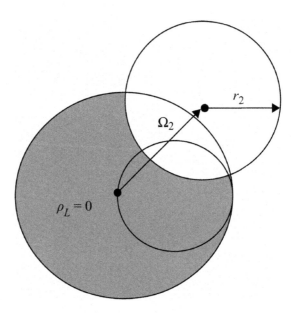

Figure 10.4. *Place of stability of* ρ_L

But

$$|S_{11}| \leq 1$$

Then

$$|S_1| < 1$$

The center of the SMITH chart is stable, and the grey part is stable too.

We can do the same if we consider the output S_2 by permuting the signs 1 and 2. We obtain the place of ρ_S, which is a circle of radius r_1 and center Ω_2 so that:

$$\Omega_1 = \frac{\left(S_{11} - \Delta S_{22}^*\right)^*}{\left|S_{11}\right|^2 - \left|\Delta\right|^2} \qquad r_1 = \frac{\left|S_{12}\right|\left|S_{21}\right|}{\left|S_{11}\right|^2 - \left|\Delta\right|^2}$$

Then we get the place stability of ρ_S (the grey part in Figure 10.5).

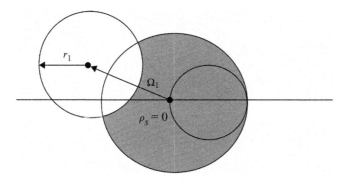

Figure 10.5. *Place of stability of* ρ_S

The different stabilities are given only for one frequency ω_1. In the case of a band of frequencies $\left(\omega_1, \omega_2, \omega_3, ..., \omega_n\right)$, we obtain a stability envelop (Figure 10.6).

10.5. Power adaptation in the case of an unconditional stability

The maximum power is going to the load (Figure 10.7) when:

$$\begin{cases} S_1 = \rho_S^* \\ S_2 = \rho_L^* \end{cases}$$

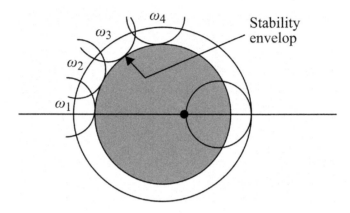

Figure 10.6. *Stability envelop*

Then, we must satisfy:

$$
\begin{cases}
S_1 = S_{11} + \dfrac{S_{12}\, S_{21}\, \rho_L}{1 - S_{22}\, \rho_L} = \dfrac{S_{11} - \Delta\, \rho_L}{1 - S_{22}\, \rho_L} \equiv \rho_S^* \\[4mm]
S_2 = S_{22} + \dfrac{S_{12}\, S_{21}\, \rho_S}{1 - S_{11}\, \rho_S} = \dfrac{S_{22} - \Delta\, \rho_S}{1 - S_{11}\, \rho_S} \equiv \rho_L^*
\end{cases}
$$

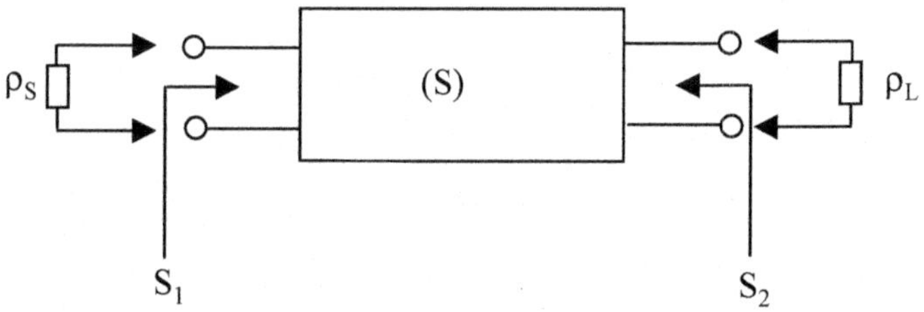

Figure 10.7. *Power adaptation*

Taking the conjugate of the first equation, we get:

$$
S_1^* = \rho_S = \left(\frac{s_{11} - \Delta\, \rho_L}{1 - S_{22}\, \rho_L} \right)^* = \frac{S_{11}^* - \Delta^*\, \rho_L^*}{1 - S_{22}^*\, \rho_L^*}
$$

and using the second equation, we get:

$$\rho_S = \frac{S_{11}^* - \Delta^* \dfrac{S_{22} - \Delta \rho_S}{1 - S_{11} \rho_S}}{1 - S_{22}^* \dfrac{S_{22} - \Delta \rho_S}{1 - S_{11} \rho_S}}$$

or

$$\rho_S = \frac{S_{11}^* \left(1 - S_{11} \rho_S\right) - \Delta^* \left(S_{22} - \Delta \rho_S\right)}{\left(1 - S_{11} \rho_S\right) - S_{22}^* \left(S_{22} - \Delta \rho_S\right)}$$

This can be written as:

$$\rho_S = \frac{\rho_S \left(|\Delta|^2 - |S_{11}|^2\right) + \left(S_{11}^* - \Delta^* S_{22}\right)}{\rho_S \left(\Delta S_{22}^* - S_{11}\right) + \left(1 - |S_{22}|^2\right)}$$

i.e.

$$\left\{ \begin{array}{l} \rho_S^2 \overbrace{\left(S_{11} - \Delta S_{22}^*\right)}^{C_1} - \rho_S \overbrace{\left(1 + |S_{11}|^2 - |S_{22}|^2 - |\Delta|^2\right)}^{B_1} + \overbrace{\left(S_{11}^* - \Delta^* S_{22}\right)}^{C_1^*} = 0 \\ \qquad\text{permuting 1 and 2 and also } S \text{ and } L \\ \rho_L^2 \underbrace{\left(S_{22} - \Delta S_{11}^*\right)}_{C_2} - \rho_L \underbrace{\left(1 + |S_{22}|^2 - |S_{11}|^2 - |\Delta|^2\right)}_{B_2} + \underbrace{\left(S_{22}^* - \Delta^* S_{11}\right)}_{C_2^*} = 0 \end{array} \right.$$

This is of the form:

$$\left\{ \begin{array}{l} C_1 \rho_S^2 - B_1 \rho_S + C_1^* = 0 \\ C_2 \rho_L^2 - B_2 \rho_L + C_2^* = 0 \end{array} \right.$$

We can note that the modulus of the product of the roots is always unity because:

$$\left|\frac{C_1}{C_1^*}\right| = \left|\frac{C_2}{C_2^*}\right| = 1$$

We can also note that the discriminator of, for example, the first equation is written as:

$$\delta_S = B_1^2 - 4\,C_1\,C_1^* = \left(1 + |S_{11}|^2 - |S_{22}|^2 - |\Delta|^2\right)^2$$
$$- 4\left(S_{11} - \Delta\,S_{22}^*\right)\left(S_{11}^* - \Delta^*\,S_{22}\right)$$

that is

$$\delta_S = \left(1 - |S_{11}|^2 - |S_{22}|^2 + |\Delta|^2 + 2|S_{11}|^2 - 2|\Delta|^2\right)^2$$
$$- 4\left(S_{11} - \Delta\,S_{22}^*\right)\left(S_{11}^* - \Delta^*\,S_{22}\right)$$

or

$$\delta_S = \left(1 - |S_{11}|^2 - |S_{22}|^2 + |\Delta|^2\right)^2 - 4|S_{12}|^2\,|S_{21}|^2$$

$$\delta_S = 4|S_{12}|^2\,|S_{21}|^2\left\{\frac{\left(1 - |S_{11}|^2 - |S_{22}|^2 + |\Delta|^2\right)^2}{4|S_{12}|^2\,|S_{21}|^2} - 1\right\}$$

δ_S is the same by changing 1 to 2. Then the discriminator δ is the same for the source and load:

$$\delta_S = \delta_L = \delta = 4|S_{12}|^2\,|S_{21}|^2\left(K^2 - 1\right)$$

Unconditional stability imposes the ROLLET coefficient K to be greater than unity and then δ is always positive.

Unconditional stability also imposes:

$$\text{input} \begin{cases} 1-|S_{22}|^2 -|S_{12}||S_{21}|>0 \text{ or } 1-|S_{22}|^2 >|S_{12}\,S_{21}| \\ 1-|S_{11}|^2 -|S_{12}||S_{21}|>0 \text{ and } 1-|S_{11}|^2 >|S_{12}\,S_{21}| \end{cases}$$
$$\text{load}$$

Then

$$\begin{cases} B_1 =1-|\varDelta|^2 +|S_{11}|^2 -|S_{22}|^2 \text{ with the first equation} \quad B_1 >-|\varDelta|^2 +|S_{11}|^2 +|S_{12}\,S_{21}| \\ B_2 =\bar{1}-|\varDelta|^2 +|S_{11}|^2 -\overline{|S_{22}|}^2 \text{ with the first equation} \quad B_2 >-|\varDelta|^2 +|S_{22}|^2 +|S_{12}\,S_{21}| \end{cases}$$

and

$$B_1 > |S_{11}|^2 + |S_{12}\cdot S_{21}| - \overbrace{|S_{11}\,S_{22} - S_{12}\,S_{21}|^2}^{\substack{\text{small of order 4} \\ \text{with } |S_{ij}| <1}} >0$$

$$B_2 > |S_{22}|^2 + |S_{12}\cdot S_{21}| - \underbrace{|S_{11}\,S_{22} - S_{12}\,S_{21}|^2}_{\substack{\text{small of order 4} \\ \text{with } |S_{ij}| <1}} >0$$

The optimum input $\rho_{S\,\mathrm{opt}}$ and output $\rho_{L\,\mathrm{opt}}$, which give the maximum value of the power, are written as:

$$\rho_{S\,\mathrm{opt}} = \frac{B_1 - 2|S_{12}\cdot S_{21}|\sqrt{K^2 -1}}{2\,C_1}$$

$$\rho_{L\,\mathrm{opt}} = \frac{B_2 - 2|S_{12}\cdot S_{21}|\sqrt{K^2 -1}}{2\,C_2}$$

The amplifier gives the maximum power with these optimum values of ρ_S and ρ_L. It is also unconditionally stable at the input and output.

It is possible to show that the corresponding maximum gain is of the form:

$$G_{\mathrm{MAX}} = \left|\frac{S_{21}}{S_{12}}\right| \left(K - \sqrt{K^2 -1}\right)$$

This maximum power gain G_{MAX} depends only on the ROLLET coefficient, $|S_{21}|$ and $|S_{12}|$.

It is a particularity of the transistor.

10.6. Bibliography

[BAH 03] BAHL I.J., BARTHIA P., *Microwave Solid State Circuit Design*, Wiley, 2003.

[BAH 09] BAHL I.J., *Fundamental of RF and Microwave Transistors Amplifiers*, Wiley-Interscience, 2009.

[CHA 94] CHANG K., *Microwave Solid-State Circuits and Applications*, Wiley-Interscience, 1994.

[EDW 81] EDWARDS T.C., *Foundations for Microstrip Circuits Design*, John Wiley & Sons, 1981.

[GEN 84] GENTILI C., *Amplificateurs et Oscillateurs Microondes* Masson, 1984.

[HA 81] HA T.T., *Solid-State Microwave Amplifier Design*, Wiley-Interscience, 1981.

[JAR 04] JARRY P., Circuits Actifs Microondes: Amplificateurs, Oscillateurs, University of Bordeaux, 2004.

[PEN 88] PENNOCK S.R., SHEPHERD P.R., *Microwave Engineering with Wireless Applications*, McGraw-Hill Telecommunications, 1988.

[SOA 88] SOARES R., *GaAs MESFET Circuit Design*, Artech House, 1988.

Noise

11.1. Introduction

We have to give a model of the noise factor F of a network. The performance of an amplifier needs the active noise components of the transistor (field effect transistor (FET) in general) to know the noise of the input and output matching circuit. The matching circuits are in general used to minimize the noise of the amplifier.

In general the input matching minimizes the noise while the output matching optimizes the gain, the power, etc.

11.2. Sources of noise

We consider a network (N) with intern sources (of noise).

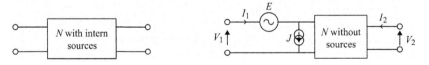

Figure 11.1. *Network with and without sources (of noise)*

These sources $(E \text{ and } J)$ have been shifted to the input and the new network is characterized by its $(ABCD)$ matrix.

$$\begin{cases} V_1 = AV_2 - BI_2 + E \\ I_1 = CV_2 - DI_2 + J \end{cases}$$

E is a tension generator of noise and J is a current generator of noise.

Now suppose we have a generator with an intern and imaginary admittance:

$$Y = G + jB$$

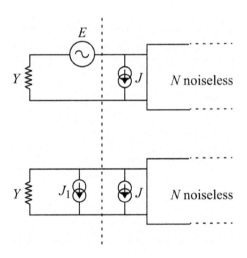

Figure 11.2. *Series and parallel noise generator*

From Figure 11.2, the series source of noise E can be replaced by a parallel source of noise J_1. Then, by definition we have:

$$J_1 = EY = E(G + jB) = J_1' + J_1''$$

Frequently, the same physique origin is at the basis of a part of the noise sources J_1 and J. Then, there is a correlation between parts of J_1 and J.

We divide J:

– into a part correlated to J_1;

– into a part non-correlated to J_1.

$$J = J_{NC} + J_C$$

J_C has a part which is in phase with J and a part which is in quadrate with J. There is also correlation admittance $Y_C = G_C + jB_C$:

$$J_C = J_C' + jJ_C'' = EY_C = E(G_C + jB_C)$$

If we recall, we have the different noise currents as shown in Figure 11.3.

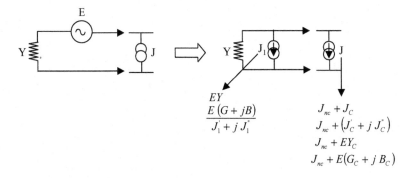

Figure 11.3. *The correlations of the noise generators*

The scalar product of J by E^* is:

$$E^* J = E^* J_{NC} + \overline{E}^2 Y_C$$

But E and J_{NC} are not correlated, then $E^*J = 0$ and:

$$Y_C = \frac{E^*J}{\overline{E}^2}$$

where $\overline{E}^2$ is the spectral noise density of the source E.

The total equivalent source of correlated noise is written as:

$$J_{TC} = E(G + jB) + E(G_C + jB_C)$$

But the non-correlated noise is J_{NC}.

And the available power of these sources connected to the load admittance $Y = G + jB$:

$$P_T = P_{CORRELATED} + P_{NON-CORRELATED}$$

which is:

$$P_T(Y) = \frac{\overline{E}^2\left\{(G+G_C)^2 + (B+B_C)^2\right\} + \overline{J}^2_{NC}}{4G}$$

We refer to R_n as the equivalent resistance of noise and G_n as the equivalent conductance of noise, and the power spectral densities of the sources of noises E and J_{NC} are:

$$\overline{E}^2 = 4kT_0 R_n \text{ / hertz} \qquad k : \text{Boltzmann's constant}$$
$$\overline{J}^2_{NC} = 4kT_0 G_n \text{ / hertz} \qquad T_0 : \text{temperature by hertz}$$

and

$$P_T \text{ / hertz} = \frac{4kT_0 R_n}{4G}\left\{(G+G_C)^2 + (B+B_C)^2\right\} + \frac{4kT_0 G_n}{4G}$$

which is:

$$P_T \,/\, \text{hertz} = k\,T_0 \left\{ \frac{R_n}{G} \left| y + y_C \right|^2 + \frac{G_n}{G} \right\}$$

11.3. Noise factor

In the case of a transistor, we define F as the noise factor:

$$F = 1 + \frac{P_T}{k\,T_0 \,/\, \text{hertz}}$$

Then

$$F = 1 + \left[\frac{R_n}{G} \left| y + y_C \right|^2 + \frac{G_n}{G} \right]$$

This quantity is important because the FET is defined by its characteristic constants as R_n, G_n and Y_C.

The real parts and the imaginary parts of Y and Y_C are written as:

$$\begin{cases} Y = G + j\,B \\ Y_C = G_C + j\,B_C \end{cases}$$

The noise factor can be written as:

$$F = 1 + R_n \frac{\left(G + G_C\right)^2 + \left(B + B_C\right)^2}{G} + \frac{G_n}{G}$$

This quantity is minimum if:

$$B_{\text{OPT}} = -\,B_C$$

and

$$G_{\text{OPT}} = \sqrt{G_c^2 + \frac{G_n}{R_n}}$$

Then, we have:

$$F_{\min} = 1 + \frac{R_n}{G_{\text{OPT}}}\left(G_{\text{OPT}} + G_C\right)^2 + \frac{G_n}{G_{\text{OPT}}}$$

or

$$F_{\min} = 1 + \frac{R_n}{G_{\text{OPT}}}\left[\left(G_{\text{OPT}} + G_C\right)^2 + \left(G_{\text{OPT}}^2 - G_C^2\right)\right]$$

which is:

$$F_{\min} = 1 + \frac{2R_n}{G_{\text{OPT}}}\left[G_{\text{OPT}}^2 + G_{\text{OPT}} \cdot G_C\right]$$

$$F_{\min} = 1 + 2\,R_n\left[G_{\text{OPT}} + G_C\right]$$

and after computing, we get:

$$F = F_{\min} + \frac{R_n}{G}\left[\left(G - G_{\text{OPT}}\right)^2 + \left(B - B_{\text{OPT}}\right)^2\right]$$

The noise factor is entirely defined by the following four characteristic parameters:

$$F_{\min}, R_n, G_{\text{OPT}}, B_{\text{OPT}}$$

and also closed by the access $Y = G + jB$.

The same is followed with the resistances and we get the result:

$$F = F_{min} + \frac{G_n}{R}\left[(R - R_{OPT})^2 + (X - X_{OPT})^2\right]$$

Then the noise factor is entirely defined by the following characteristic parameters:

$$F_{min}, G_n, R_{OPT}, X_{OPT}$$

11.4. Noise circles

We propose to give a representation of the noise on the Smith chart.

Remember, we had:

$$F = F_{min} + \frac{G_n}{R}\left[(R - R_{OPT})^2 + (X - X_{OPT})^2\right]$$

This means that the next quantity C is a constant:

$$C^{nt} = C = \frac{F - F_{min}}{G_n} = \frac{1}{R}\left[(R - R_{OPT})^2 + (X - X_{OPT})^2\right]$$

$$CR = (R - R_{OPT})^2 + (X - X_{OPT})^2$$

Now, we have:

$$CR = (R - R_{OPT})^2 + (X - X_{OPT})^2$$

with

$$R_{OPT} = R_A + X \; ; X_{OPT} = X_A$$

Then, we get circles equations that are called the noise circles:

$$(R - R_A)^2 + (X - X_A)^2 = A^2$$

with

$$\begin{cases} R_A = R_{OPT} + \dfrac{C}{2} \\ X_A = X_{OPT} \\ A^2 = C\,R_{OPT} + \dfrac{C^2}{4} \end{cases}$$

When we have the noise factor F, then C, R_A, A^2 are obtained.

Then, we get the noise circles on the Smith chart.

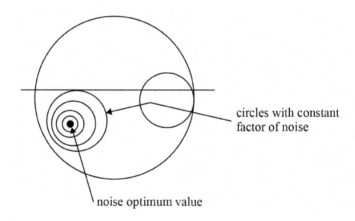

circles with constant factor of noise

noise optimum value

Figure 11.4. *Noise circles*

11.5. Bibliography

[BAH 09] BAHL I.J. *Fundamental of RF and Microwave Transistors Amplifiers*, Wiley-Interscience, 2009.

[EDW 81] EDWARDS T.C., *Foundations for Microstrip Circuits Design*, John Wiley & Sons, 1981.

[FRI 44] FRIIS H.T., "Noise figures of radio receivers" *Proceedings of IRE*, vol. 32, pp. 419–422, July 1944.

[GEN 84] GENTILI C., *Amplificateurs et Oscillateurs Microondes*, Masson, 1984.

[HA 81] HA T.T., *Solid-State Microwave Amplifier Design*, Wiley-Interscience, 1981.

[JAR 04] JARRY P., Circuits Actifs Microondes: Amplificateurs, Oscillateurs, University of Bordeaux, 2004.

[PEN 88] PENNOCK S.R., SHEPHERD P.R., *Microwave Engineering with Wireless Applications*, McGraw-Hill Telecommunications, 1988.

[SOA 88] SOARES R., *GaAs MESFET Circuit Design*, Artech House, 1988.

[VAN 86] VAN DER ZIEL A., *Noise in Solid State Devices and Circuits*, Wiley-Interscience, 1986.

Problems

12.1. Symmetric writing of G_T in the case of the non-unilateral amplifier

Remember the power gain is written as:

$$G_T = \frac{\left(1-|\rho_s|\right)^2 |S_{21}|^2 \left(1-|\rho_L|^2\right)}{\left|\left(1-S_{11}\,\rho_S\right)\left(1-S_{22}\,\rho_L\right)-S_{12}\,S_{21}\,\rho_S\,\rho_L\right|^2}$$

And we want this G_T to be written in the same form as G_{TU} :

$$\begin{cases} G_T = \dfrac{\left(1-|\rho_S|^2\right)|S_{21}|^2\left(1-|\rho_L|^2\right)}{\left|\left(1-S_1\,\rho_S\right)\left(1-S_2\,\rho_L\right)^2\right|} \\[4mm] \text{to be compared with } G_{T\,u}\ \left(S_{12}=0\right) \\[2mm] G_{Tu} = \dfrac{\left(1-|\rho_S|^2\right)|S_{21}|^2\left(1-|\rho_L|^2\right)}{\left|\left(1-S_{11}\,\rho_S\right)\left(1-S_{22}\,\rho_L\right)^2\right|} \end{cases}$$

1) To do this, compute the reflection $S_1 = b_1/a_1$ coefficient when the network is closed on the load ρ_L .

2) Deduce the expression of the power gain.

The solution

1) We have the following flow graph:

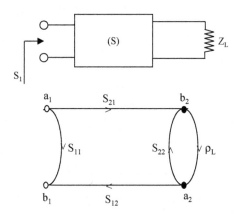

And we use Masson's non-touching rule to compute the report
$S_1 = \dfrac{b_1}{a_1}$.

We go from a_1 to b_1 and the gains are $T_1 = S_{11}$ and $T_2 = S_{21} \, \rho_L \, S_{12}$.

$\Delta = 1 -$ (sum of all the first-order loops) + (sum of all the second-order loops)

Then:

$$\Delta = 1 - S_{22} \, \rho_L$$

and:

$\Delta_1 = 1 -$ (sum of the first-order loop without touching T_1) + (sum of the second-order loop without touching T_1)

$$\Delta_1 = 1 - S_{22}\,\rho_L$$

$\Delta_2 = 1 -$ (sum of the first-order loop without touching T_2) + (sum of the second-order loop without touching T_2)

$$\Delta_2 = 1$$

Then:

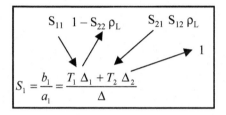

or:

$$S_1 = \frac{S_{11}\left(1 - S_{22}\rho_L\right) + S_{12}\,S_{21}\,\rho_L}{1 - S_{22}\,\rho_L}$$

This gives:

$$S_1 = S_{11} + \frac{S_{12}\,S_{21}\,\rho_L}{1 - S_{22}\,\rho_L}$$

This has been done for the input; we can also do for the output by using the changes:

$$1 <\!\text{---------}\!> 2$$

$$L <\!\text{---------}\!> S$$

This gives:

$$S_2 = S_{22} + \frac{S_{12}\,S_{21}\,\rho_S}{1 - S_{11}\,\rho_S}$$

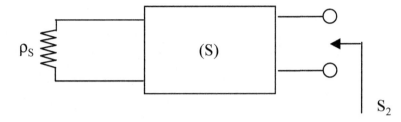

2) Remember G_T is written as:

$$G_T = \left(1-|\rho_L|^2\right)\left(1-|\rho_S|^2\right)\left|\frac{S_{21}}{\left(1-S_{11}\,\rho_S\right)\left(1-S_{22}\,\rho_L\right)-S_{12}\,S_{21}\,\rho_S\,\rho_L}\right|^2$$

$$G_T = \left(1-|\rho_L|^2\right)\left(1-|\rho_S|^2\right)\left|\frac{b_2}{b_S}\right|^2$$

Dividing by $(1 - S_{22}\,\rho_L)$, we get:

$$\frac{b_2}{b_S} = \frac{S_{21}/\left(1-S_{22}\,\rho_L\right)}{1-S_{11}\,\rho_S - \dfrac{S_{11}\,S_{21}\,\rho_L}{1-S_{22}\,\rho_L}\,\rho_S}$$

or:

$$\frac{b_2}{b_s} = \frac{S_{21}/\left(1-S_{22}\,\rho_L\right)}{1-S_1\,\rho_S}$$

Which means that:

$$\frac{b_2}{b_s} = \frac{S_{21}}{\left(1-S_1\,\rho_S\right)\left(1-S_{22}\,\rho_L\right)}$$

And a symmetrical expression:

$$G_T = \frac{\left(1-|\rho_S|^2\right)|S_{21}|^2\left(1-|\rho_L|^2\right)}{\left|1-S_1\,\rho_S\right|^2\left|1-S_{22}\,\rho_L\right|^2}$$

By the same manner, we also get:

$$G_T = \frac{\left(1-|\rho_S|^2\right)|S_{21}|^2\left(1-|\rho_L|^2\right)}{|1-S_{11}\,\rho_S|^2\,|1-S_2\,\rho_L|^2}$$

The power gain G_T has the same form as in the unilateral case.

12.2. Stability conditions of a broadband transistor from 300 to 900 MHz

We give the scattering parameters (S_{ij}) of the following transistor from 300 to 900 MHz:

MHz	S_{11}	S_{22}	S_{21}	S_{12}
300	$0.267\angle-88.5°$	$0.57\angle-30.8°$	$11.61\angle112°$	$0.043\angle68°$
500	$0.178\angle-122°$	$0.52\angle-29°$	$7.37\angle99°$	$0.060\angle70°$
700	$0.122\angle-156°$	$0.49\angle-28°$	$5.45\angle89°$	$0.082\angle69°$
900	$0.111\angle-179°$	$0.46\angle-27°$	$3.75\angle77°$	$0.106\angle69°$

1) What are the stabilities conditions at the different frequencies (300 MHz, 500 MHz, 700 MHz and 900 MHz)?

2) Gives the curve of the factor of Rollet K with the frequency. Can you give a Conclusion?

The solution

1) We have to verify the three conditions for all the frequencies:

$$\begin{cases} K = \dfrac{1+|\Delta|^2-|S_{11}|^2-|S_{22}|^2}{2\,|S_{12}\,S_{21}|} > 1 \ (\text{ROLLET}) \\ 1-|S_{22}|^2-|S_{12}\,S_{21}| > 0 \ (S_1 \text{ condition}) \\ \text{and} \\ 1-|S_{11}|^2-|S_{12}\,S_{21}| > 0 \ (S_2 \text{ condition}) \end{cases}$$

300 MHz: we compute first: $\Delta = S_{11}S_{22} - S_{12}S_{21}$. We find:

$$\Delta = 0.428 + j0.128 = 0.447e^{j0.308\,Rd} = 0.447\angle 17.55°$$
$$|\Delta| = 0.447$$
$$|\Delta|^2 = 0.20$$

and:

$$\begin{cases} K = \dfrac{1 + 0.2 - (0.267)^2 - (0.57)^2}{2*11.61*0.043} = 0.805 \le 1\,(Rollet) \\ \text{but} \\ 1 - |S_{22}|^2 - |S_{12}\,S_{21}| = 0.584 \ge 0 \\ 1 - |S_{11}|^2 - |S_{12}\,S_{21}| = 0.753 \ge 0 \end{cases}$$

The coefficient of Rollet is less than unity and the transistor is unstable at 300 MHz.

500 MHz: the transistor is still unstable because $K = 0.95$ and is less than unity. But the two other conditions are satisfied.

700 MHz: in this case, we have:

$$\Delta = 0.3503 - j0.162 = 0.386e^{-j0.498\,Rd} = 0.386\angle -22.11°$$
$$|\Delta| = 0.386$$
$$|\Delta|^2 = 0.149$$

and:

$$\begin{cases} K = \dfrac{1 + 0.149 - (0.122)^2 - (0.49)^2}{2*5.45*0.0082} = 1.00022 \ge 1\,(Rollet) \\ \text{and} \\ 1 - |S_{22}|^2 - |S_{12}\,S_{21}| = 0.31 \ge 0 \\ 1 - |S_{11}|^2 - |S_{12}\,S_{21}| = 0.53 \ge 0 \end{cases}$$

The coefficient of Rollet is greater than unity and the transistor is unconditionally stable at 700 MHz.

900 MHz: the transistor is still unconditionally stable because $K = 1.28$ and is greater than unity. And the two other conditions are satisfied.

2) We give the curve with the values of K as a function of the frequency.

MHz	K	
300	0.85	*unstable*
500	0.95	*unstable*
700	1.0002	*stable*
900	1.128	*stable*

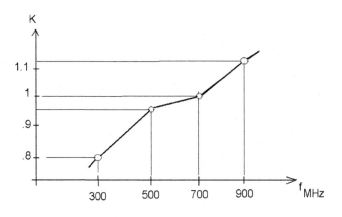

The transistor is unconditionally stable after 700 MHz.

12.3. Narrow band amplifier around 500 MHz

1) The scattering parameters $\left(S_{ij}\right)$ of a microwave transistor have been measured at $f_0 = 500$ MHz and at $f_1 = 550$ MHz.

	500 MHz	550 MHz
S_{11}	$0.343\angle-174°$	$0.345\angle-177°$
S_{21}	$6.311\angle85°$	$5.774\angle82°$
S_{12}	$0.058\angle72°$	$0.063\angle72°$
S_{22}	$0.441\angle-23°$	$0.390\angle-21°$

Show that the transistor is stable at $f_0 = 500$ MHz.

2) Show that this transistor is still stable at $f_1 = 550$ MHz.

3) What is the maximum unilateral gain at $f_1 = 550$ MHz?

4) What is the maximum power gain at $f_1 = 550$ MHz?

Now, we place input and output matching circuits of order 2 (one inductance and one capacity).

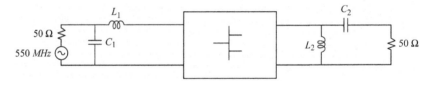

5) Give the values of the input elements C_1 and L_1.

6) Give the values of the output elements C_2 and L_2.

7) Give the response of this amplifier.

The solution

1) As in the preceding problem, we have: $\Delta = S_{11}S_{22} - S_{12}S_{21}$

$$|\Delta| = 0.216$$
$$|\Delta|^2 = 0.0466$$
$$K = 1.0035 \geq 1$$

Moreover:

$$1-\left|S_{22}\right|^2-\left|S_{12}\,S_{21}\right|=0.44\geq0$$

$$1-\left|S_{11}\right|^2-\left|S_{12}\,S_{21}\right|=0.516\geq0$$

The three are verified and the transistor is unconditionally stable at $f_0 = 500$ MHz.

2) The modulus of S_{21} is decreasing with the frequency and it is necessary to design the amplifier with a slightly larger frequency. We choose a frequency of $f_1 = 550$ MHz.

We have: $\Delta_e = S_{11e}S_{22e} - S_{12e}S_{21e}$

$$\left|\Delta_e\right| = 0.2312$$

$$\left|\Delta_e\right|^2 = 0.05345$$

$$K_e = 1.0754 \geq 1$$

and:

$$1-\left|S_{22e}\right|^2-\left|S_{12e}\,S_{21e}\right|=0.484\geq0$$

$$1-\left|S_{11e}\right|^2-\left|S_{12e}\,S_{21e}\right|=0.516\geq0$$

The transistor is unconditionally stable at $f_1 = 550$ MHz.

3) The maximum of the unilateral gain $\left(S_{12}=0\right)$ is given after perfect matching at the input $\rho_S = S_{11e}^*$ and at the output $\rho_L = S_{22e}^*$.

$$G_{Tu\,\max} = \frac{1}{1-\left|S_{11e}\right|^2}\left|S_{21e}\right|^2\frac{1}{1-\left|S_{22e}\right|^2}$$

where $G_S = \dfrac{1}{1-|S_{11e}|^2} = \dfrac{1}{0.885} = 1.135$ is the maximum unilateral gain of the input.

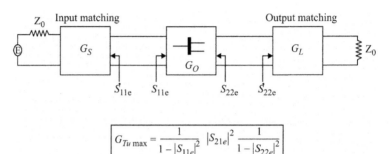

$$G_{Tu\,max} = \frac{1}{1-|S_{11e}|^2}\,|S_{21e}|^2\,\frac{1}{1-|S_{22e}|^2}$$

In decibels, $G_{Sdb} = 10\log 1.135 = 0.55$ db .

Also, the intrinsic gain $G_0 = |S_{21e}|^2 = 33.34$.

In decibels, $G_{0db} = 10\log 33.34 = 15.25$ db.

And at the output, $G_L = \dfrac{1}{1-|S_{22e}|^2} = \dfrac{1}{0.8479} = 1.179$.

In decibels, $G_{Ldb} = 10\log 1.179 = 0.72$ db.

In considering the whole chain:

$$G_{Tu\,max} = G_S G_O G_L$$

And in decibels:

$$G_{Tu\,max\,db} = G_{Sdb} + G_{Odb} + G_{Ldb}$$

We have:

$$G_{Tu\,max\,db} = 16.5\,db$$

4) The maximum power gain at $f_1 = 550$ MHz is:

$$G_{max} = \frac{|S_{21e}|}{|S_{12e}|}\left(K_e - \sqrt{K_e^2 - 1}\right)$$

$$G_{max} = \frac{5.774}{0.063}\left(1.0754 - \sqrt{(1.0754)^2 - 1}\right)$$

and:

$$G_{max} = 62.30$$
$$G_{max\,db} = 10\log 62.30 = 17.945 \text{ db}$$

The error with the maximum unilateral gain is:

$$17.945 \text{ db} - 16.5 \text{ db} = 1.445 \text{ db}$$

5) The maximum power matching is given at the input when:

$$\begin{cases} S_{1e} = \rho_{Se}^* \\ S_{2e} = \rho_{Le}^* \end{cases}$$

$$S_{1e} \qquad\qquad\qquad S_{2e}$$

Then, we have at the input:

$$\begin{cases} \rho_{SOPTe} = \dfrac{B_{1e} - 2|S_{12e}S_{21e}|\sqrt{K_e^2 - 1}}{2C_{1e}} \\ \text{and} \\ B_{1e} = 1 - |A_e|^2 + |S_{11e}|^2 - |S_{22e}|^2 \\ C_{1e} = S_{11e} - A_e S_{22e}^* \end{cases}$$

And after computation, we get:

$$\rho_{SOPTe} = 0.7231\angle 180° = -0.7231$$

Synthesis of the input matching network needs to know the low-pass elements of the next circuit.

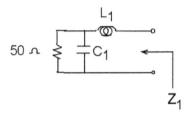

with:

$$\rho_{SOPTe} = \frac{Z_1 - 50}{Z_1 + 50}$$

and:

$$Z_1 = 50\frac{1 + \rho_{SOPTe}}{1 - \rho_{SOPTe}} = 8.1$$

but:

$$Z_1 = \frac{1}{\dfrac{1}{50} + jC_1\omega} + jL_1\omega$$

$$Z_1 = \frac{50}{1 + \left(50C_1\omega\right)^2} + j\omega\left\{ L_1 - \frac{2500C_1}{1 + \left(50C_1\omega\right)^2} \right\} \cong 8.1$$

The real part is equal to 8.1 but the imaginary part is zero.

It first gives:

$$
\begin{cases}
C_1\omega = \dfrac{1}{50}\sqrt{\dfrac{50}{8.1}-1} \\[3mm]
L_1 = \dfrac{2500\,C_1}{1+\left(50\,C_1\omega\right)^2} \\[3mm]
\omega = 2\pi f
\end{cases}
$$

and:

$$
\begin{cases}
C_1 = 13.2\,\mathrm{pF} \\
L_1 = 5.3\,\mathrm{nH} \\
at\ f = 550\,\mathrm{MHz}
\end{cases}
$$

6) Then, we have at the output:

$$
\begin{cases}
\rho_{LOPTe} = \dfrac{B_{2e} - 2\left|S_{12e}S_{21e}\right|\sqrt{K_e^2 - 1}}{2C_{2e}} \\[3mm]
\text{and} \\[2mm]
B_{2e} = 1 - \left|\Delta_e\right|^2 - \left|S_{11e}\right|^2 + \left|S_{22e}\right|^2 \\
C_{2e} = S_{22e} - \Delta_e S_{11e}^*
\end{cases}
$$

And after computation, we get:

$$\rho_{LOPTe} = 0.7386 \angle 23^\circ = 0.680 + j0.289$$

Synthesis of the input matching network needs to know the high-pass elements of the circuit.

$$
\begin{cases}
Z_2 = 50\dfrac{1+\rho_{LOPTe}}{1-\rho_{LOPTe}} = 122.04 + j155.40 \\[3mm]
Y_2 = 3.126*10^{-3} - j3.98*10^{-3}
\end{cases}
$$

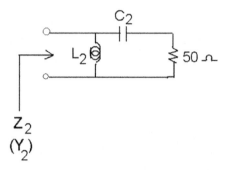

Z_2
(Y_2)

And we must have:

$$\begin{cases} Y_2 = \dfrac{1}{j\omega L_2} + \dfrac{1}{\dfrac{1}{j\omega C_2} + 50} \\[4mm] Y_2 = \dfrac{50}{(50)^2 + X^2} - j\dfrac{50^2 B - X(1 - BX)}{(50)^2 + X^2} \cong 3.126 * 10^{-3} - j3.98 * 10^{-3} \\[4mm] \text{with} \\[2mm] B = \dfrac{1}{L_2 \omega} \quad \text{and} \quad X = \dfrac{1}{C_2 \omega} \end{cases}$$

After identification and computation, we arrive at:

$$\begin{cases} C_2 = 2.49 \text{ pF} \\ L_2 = 25.8 \text{ nH} \\ at \ f = 550 \text{ MHz} \end{cases}$$

7) Using the computed values of the capacities (C_1, C_2) and of the inductances (L_1, L_2), we get perfect matching at the input and the output and an amplification of 17.945 db at 550 MHz that corresponds to a maximum transfer of power.

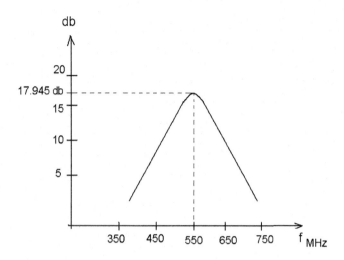

12.4. Low-noise amplifier at 2.5 GHz

We propose to use a microwave transistor at $f_0 = 2.5$ GHz with the measured scattering parameters S_{ij}.

	$f_0 = 2.5$ GHz
S_{11}	$0.570 \angle -163°$
S_{22}	$0.760 \angle -50°$
S_{21}	$2.646 \angle 59°$
S_{12}	$0.042 \angle 25°$

We also give the parameters of noise at the same frequency $f_0 = 2.5$ GHz:

$$\begin{cases} F_{min} = 1.8 \text{ db} \\ R_n = 13.8 \text{ } \Omega \\ \Gamma_{OPT} = 0.412 \angle 104° \end{cases}$$

where R_n is the equivalent resistance of noise and Γ_{OPT} the optimum reflection coefficient that gives the minimum value of the noise $F_{\min}$.

1) Check the stability at $f_0 = 2.5$ GHz.

2) What is the maximum transfer power gain at $f_0 = 2.5$ GHz?

3) What is the reflection coefficient optimum in the side of the source ρ_{SOPT}?

4) Give the value of the factor of noise at $f_0 = 2.5$ GHz.

The solution

1) We have:

$$\Delta = S_{11}S_{22} - S_{12}S_{21}$$

And with the values of the scattering parameters S_{ij}:

$$\Delta = -0.374 + j0.124 = 0.394 \angle 162°$$
$$|\Delta| = 0.394$$
$$|\Delta|^2 = 0.155$$

Then, the Rollet factor is:

$$K = \frac{1 + |\Delta|^2 - |S_{11}|^2 - |S_{22}|^2}{2|S_{12}S_{21}|}$$

i.e.:

$$K = \frac{1 + 0.155 - 0.325 - 0.577}{2 * 0.111} = 1.14$$

This quantity is more important than unity.

Moreover, the two other conditions are verified:

$$1 - \left| S_{22} \right|^2 - \left| S_{12} S_{21} \right| \geq 0$$
$$1 - \left| S_{11} \right|^2 - \left| S_{12} S_{21} \right| \geq 0$$

And the transistor is unconditionally stable at $f_0 = 2.5 \text{ GHz}$.

2) The maximum transfer power gain at $f_0 = 2.5 \text{ GHz}$ is given using the formula:

$$G_{MAX} = \left| \frac{S_{21}}{S_{12}} \right| \left(K - \sqrt{K^2 - 1} \right)$$

This corresponds to a gain:

$$G_{MAX} = \frac{2.646}{0.042} \left(1.14 - \sqrt{(1.14)^2 - 1} \right) = 37.336$$

$$G_{MAXdb} = 10 \log 37.336 = 15.7 \text{ db}$$

3) The power matching input is obtained when:

$$\begin{cases} S_1 = \rho_S^* \\ S_2 = \rho_L^* \end{cases}$$

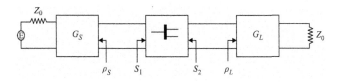

$$\begin{cases} \rho_{SOPT} = \dfrac{B_1 - 2 \left| S_{12} S_{21} \right| \sqrt{K^2 - 1}}{2 C_1} \\ \text{and} \\ B_1 = 1 - \left| \Delta \right|^2 + \left| S_{11} \right|^2 - \left| S_{22} \right|^2 \\ C_1 = S_{11} - \Delta S_{22}^* \end{cases}$$

In our case, a numerical application gives:

$$\begin{cases} B_1 = 1 - 0.155 + 3.25 - 0.577 = 0.593 \\ C_1 = 0.570\, e^{-j163} - 0.394 * 0.760\, e^{j162} e^{j50} = 0.288 \angle 180° \end{cases}$$

and:

$$\rho_{SOPT} = \frac{0.593 - 2 * 0.042 * 2.646 \sqrt{(1.14)^2\, 61}}{2 * 0.289} \angle -180°$$

It is:

$$\rho_{SOPT} = 0.816 \angle -180°$$

Using the same method and changing the indices $1 \leftrightarrow 2$ and $S \leftrightarrow L$, we will find:

$$\rho_{LOPT} = 0.894 \angle 56°$$

This last result is not because in the continuation of the problem, we need only the optimum reflection coefficient on the side of the source that is equal to Γ :

$$\rho_{SOPT} = \Gamma$$

4) From the next figure, the input factor of noise is F.

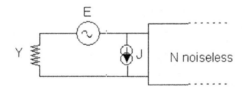

$$F = F_{min} + \frac{R_n}{G}\left[(G - G_{OPT})^2 + (B - B_{OPT})^2 \right]$$

$Y = G + jB$: input admittance of the load,

R_n : noise resistance,

G_{OPT}, B_{OPT} : optimum values of the point with a minimum noise factor F_{min}.

It can be written as:

$$F = F_{min} + \frac{R_n}{G}|Y - Y_{OPT}|^2$$

And using the reduced admittances:

$$\begin{cases} y = R_0 Y = g + jb \\ y_{OPT} = R_0 Y_{OPT} = g_{OPT} + jb_{OPT} \\ R_0 = 50\,\Omega \end{cases}$$

We get:

$$\begin{cases} F = F_{min} + \dfrac{r_n}{g}\left[(g - g_{OPT})^2 + (b - b_{OPT})^2\right] \\ F = F_{min} + \dfrac{r_n}{g}|y - y_{OPT}|^2 \end{cases}$$

Or using the reflecting coefficients:

$$\Gamma = \frac{1-y}{1+y} \text{ and } \Gamma_{OPT} = \frac{1 - y_{OPT}}{1 + y_{OPT}}$$

We find after a fastidious computation:

$$F = F_{min} + 4r_n \frac{|\Gamma - \Gamma_{OPT}|^2}{\left(1 - |\Gamma|^2\right)|1 + \Gamma_{OPT}|^2}$$

We know all the elements at $f_0 = 2.5$ GHz :

$$\begin{cases} F_{min} = 1.8 \text{ db} \\ R_n = 13.8\Omega \\ \Gamma_{OPT} = 0.412\angle 104° \end{cases}$$

And we placed in the side of the source with a Γ corresponding to a maximum power transfer:

$$\Gamma = \rho_{SOPT} = 0.816\angle -180°$$

Then:

$$F_{min\, db} = 1.8 \; db = 10\log 1.51 \Rightarrow F_{min} = 1.51$$

$$F = 1.51 + 4*0.276 \frac{\left| -0.816 - 0.412 e^{j104°}\right|^2}{\left(1 - (0.816)^2\right)\left|1 + 0.412 e^{j104°}\right|^2}$$

$$F = 1.51 + 2.29 = 3.80$$

And in decibel:

$$F_{db} = 10\log 3.80 = 5.80$$

$$F_{db} = 5.80 \text{ db}$$

And we can see that we are quite far from the minimum that is $F_{min\, db} = 1.8$ db .

12.5. Bibliography

[BAH 09] BAHL I.J., *Fundamental of RF and Microwave Transistors Amplifiers*, Wiley-Interscience, 2009.

[GEN 84] GENTILI C., *Amplificateurs et Oscillateurs Microondes*, Masson, 1984.

[JAR 98] JARRY P., KERHERVE E., Lignes Microondes Couplées, Filtres sur Guide, Amplificateurs Microondes, ENSEIRB University of Bordeaux, 1998.

PART 4

Microwave Oscillators

13

Quasi-static Analysis and Overvoltage Coefficients of an Oscillator

13.1. Introduction

The classification of microwave oscillators is based on three parameters:

- the power;

– the possibility of electric tuning;

– spectral purity.

The above three parameters are not compatible because:

– a low power with a good spectra oscillator is difficult to tune;

– a power oscillator has no pure spectra;

– A tuned oscillator has an output, which of low power.

But it is possible to obtain two of the three properties with a cascade of two particular oscillators (for example a tuned oscillator and a power oscillator give a tuned power oscillator).

We distinguish between the power oscillators (several megawatts, such as the Klystron, Magnetron, TOP, etc.) and the oscillators of medium and low power (several watts or milliwatts, such as the Gunn diodes, the bipolar oscillators, the FET oscillators, etc.).

The important characteristics are different if we consider oscillators from emission and reception. In Figure 13.1, we give the principal properties of these two kinds of oscillators.

Figure 13.1. *Properties of emission/reception oscillators*

The microwave oscillators use active semiconductor components. It is possible to analyze these oscillators by using the concept of negative and nonlinear (NL) resistance.

13.2. Quasi-static analysis of the microwave oscillators

In the microwave domain, the rise time is very important because it represents about some half a score to some hundreds of cycles of the circuit.

During the transitory time we get a current with a complex pulse.

$$i(t) = A_0\, e^{pt} \text{ with } \frac{di}{dt} = p\, i(t)$$

where A_0 and p are the modulus and the phase, which are independent of the time t.

A_0 and p are a function of the time, and variables $A_0(t)$ and $p(t)$ are the local constant and quasi-static.

$$i(t) = A(t)e^{j\varphi(t)} \text{ with } \frac{di}{dt} = \underbrace{\left[\frac{1}{A}\frac{dA}{dt} + j\frac{d\varphi}{dt} \right]}_{P} i(t)$$

Making a comparison with these two equations, we have the equivalence:

$$p = \frac{1}{A(t)}\frac{dA}{dt} + j\frac{d\varphi}{dt} = \alpha + j\omega$$

and

$$\alpha = \frac{1}{A}\frac{dA}{dt}$$

$$\omega = \frac{d\varphi}{dt}$$

where α and ω are the generalized attenuation and generalized frequency. We can note that p is a slow variable.

13.3. NL resistances

An oscillator can be defined as a parallel mesh of an NL impedance Z_{NL} and a linear load impedance Z_L (Figure 13.2).

As an example, the resistance of a diode Gunn is NL because it depends on the current I_0, which passes through this diode. We have an NL resistance $R_{NL} = R_0 + R_1 I_0^2$.

Z_{NL} is the NL impedance of the active element (for example a transistor) and Z_L is the impedance of the load (linear element).

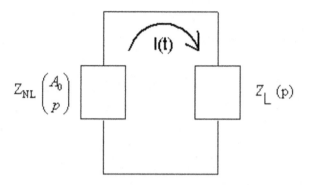

Figure 13.2. *An oscillator as a circuit*

Suppose that there is a current:

$$I(t) = A\ e^{pt}$$

where p is a slow variable with the time t.

The equation of the circuit is written as:

$$(Z_{NL} + Z_L)\ A\ e^{pt} = 0$$

But

$$A \neq 0 \text{ and } Z_{NL} = Z_{NL}\ (p, A)$$

We get the fundamental equation of the oscillators:

$$Z_T = Z_{NL}\left(p, A\right) + Z_L\left(p\right) = 0$$

For $p = \alpha + j\omega$, the equation has a real part and an imaginary part. These two parts have to be zero.

$$R_e\left\{Z_T\left(A, p\right)\right\} = 0$$
$$I_m\left\{Z_T\left(A, p\right)\right\} = 0$$

If oscillations are maintained, we have constant amplitude, constant frequency and losses, which are zero:

$$\begin{cases} A = A_0 \\ p = j\omega_0 \\ \alpha = 0 \end{cases}$$

And we arrive at the equation of the oscillators with oscillations maintained:

$$R_e\left\{Z_T\left(A_0, j\omega_0\right)\right\} = 0$$
$$I_m\left\{Z_T\left(A_0, j\omega_0\right)\right\} = 0$$

If the two impedances are real, we must have $R_{NL} + R_L = 0$ with a positive load $R_L \geq 0$.

To have oscillations, we need a negative NL resistance $R_{NL} < 0$.

13.4. Output power of the oscillator

The power is expressed is:

$$P = \frac{1}{2} R_{NL} A_0^2$$

This power has a maximum value when:

$$\frac{\partial P}{\partial A_0} = 0$$

It is given by:

$$\frac{\partial P}{\partial A_0} = \frac{1}{2} \frac{\partial R_{NL}}{\partial A_0} A_0^2 + \frac{1}{2} 2 R_{NL} A_0 \equiv 0$$

and gives:

$$A_0 \frac{\partial R_{NL}}{\partial A_0} = -2 R_{NL}$$

Knowing the curve $R_{NL} = f(A_0)$, we can deduce the optimum value $A_{0\,opt}$ and of the quantity $R_{NL\,opt}$. Then, the optimum value of the load is given by:

$$R_{L\,opt} = \left| -R_{NL\,opt} \right|$$

To get oscillations, we have to satisfy $R_{NL} + R_L = 0$ with a positive load $R_L \geq 0$.

For a Gunn diode, the equivalent circuit is shown in Figure 13.3.

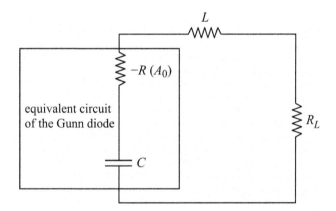

Figure 13.3. *Equivalent circuit of the Gunn diode*

The oscillating conditions can be written as:

$$\begin{cases} LC\, \omega_0^2 = 1 \\ R_L - R(A_0) = 0 \end{cases}$$

If, for example, $R(A_0) = 10\ \Omega$, we come back to $10\ \Omega$ with a transmission line of length l.

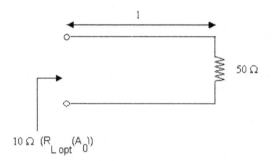

Figure 13.4. *How to see 10 Ω for $R_L(A_0)$*

13.5. Stability of the oscillation

Suppose the oscillator is working in a stable manner:

$$Z_T\left(A_0, j\omega_0\right) = 0$$

and also suppose that there are some slight perturbations and the oscillator is still working. We make a limited development of the first-order of the quantity:

$$Z_T\left(p, A\right) = Z_{NL}\left(p, A\right) + Z_L\left(p\right) = 0$$

This is:

$$Z_T\left(j\omega_0, A_0\right) + \frac{\partial Z_T}{\partial p}\, dp + \frac{\partial Z_T}{\partial A}\, dA = 0$$

The working ("evolution") equation of the oscillator can be written as:

$$\left[Z_T\left(j\omega_0, A_0\right) + \frac{\partial Z_T}{\partial p}\, dp + \frac{\partial Z_T}{\partial A}\, dA\right] A_0\, e^{pt} = 0$$

If there is still a current:

$$Z_T\left(j\omega_0, A_0\right) \equiv 0$$

We obtain the mathematical condition to maintain the oscillation:

$$\frac{\partial Z_T}{\partial p} dp + \frac{\partial Z_T}{\partial A} dA = 0$$

We can compute at $p = j\omega$:

$$\frac{\partial Z_T}{\partial p} = \frac{1}{j} \frac{\partial Z_T}{\partial \omega} = -j \frac{\partial Z_T}{\partial \omega}$$

Reporting in the maintain equation of the oscillation:

$$-j \frac{\partial Z_T}{\partial \omega} dp + \frac{\partial Z_T}{\partial A} dA = 0$$

and

$$dp = -j \frac{\dfrac{\partial Z_T}{\partial A}}{\dfrac{\partial Z_T}{\partial \omega}} dA$$

If $\left(\dfrac{\partial Z_T}{\partial \omega}\right)^*$ is the conjugate of $\left(\dfrac{\partial Z_T}{\partial \omega}\right)$:

$$dp = d\alpha + jd\omega = -j \frac{\dfrac{\partial Z_T}{\partial A} \cdot \left(\dfrac{\partial Z_T}{\partial \omega}\right)^*}{\left|\dfrac{\partial Z_T}{\partial \omega}\right|^2} dA$$

but Z_T is a complex function.

$$Z_T = R_T + jX_T \Rightarrow \begin{cases} \dfrac{\partial Z_T}{\partial \omega} = \dfrac{\partial R_T}{\partial \omega} + j\dfrac{\partial X_T}{\partial \omega} = (a + jb) \\[3mm] \dfrac{\partial Z_T}{\partial A} = \dfrac{\partial R_T}{\partial A} + j\dfrac{\partial X_T}{\partial A} = (A + jB) \end{cases}$$

Then

$$dp = d\alpha + j\,d\omega = -j\frac{(A + jB)(a - jb)}{a^2 + b^2}\,dA$$

Now, we separate the real and imaginary parts:

$$d\alpha = -\frac{Ab - Ba}{a^2 + b^2}\,dA$$

$$d\omega = -\frac{Aa + Bb}{a^2 + b^2}\,dA$$

And it results in stabilities conditions:

* when $dA > 0$ we must have $d\alpha < 0$ $\left.\vphantom{\begin{array}{c}1\\1\end{array}}\right\}$ then $\dfrac{dA}{d\alpha} < 0$

* in same way when $dA < 0$ we must have $d\alpha > 0$

which means:

$$Ab - Ba \geq 0$$

or

$$\boxed{\left.\frac{\partial R_T}{\partial A}\right|_{A_0} \cdot \left.\frac{\partial X_T}{\partial \omega}\right|_{\omega_0} - \left.\frac{\partial X_T}{\partial A}\right|_{A_0} \cdot \left.\frac{\partial R_T}{\partial \omega}\right|_{\omega_0} > 0}$$

In the same way, if we want no variation of the frequency (ω) given by a variation of the amplitude (A), in fact no conversion AM $-$ FM, we must have $\dfrac{d\omega}{dA} = 0$.

$$\frac{\partial R_T}{\partial A} \cdot \frac{\partial R_T}{\partial \omega} + \frac{\partial X_T}{\partial A} \cdot \frac{\partial X_T}{\partial \omega} = 0$$

This means that the curves $Z_T(A_0)$ given when $\omega = ct$ and the curves $Z_T(\omega)$ given when $A_0 = ct$ are orthogonal.

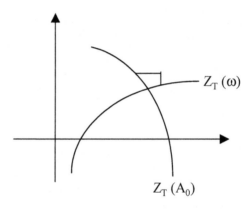

Figure 13.5. *No conversion AM–FM*

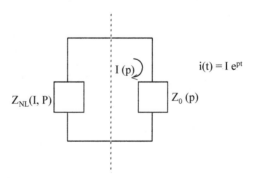

Figure 13.6. *A circuit as a general network*

13.6. Overvoltage coefficients of a microwave oscillator

13.6.1. *The case of a linear circuit*

We consider linear impedance $Z_0 = R_0 + jX_0$ with a current $I_0 e^{j\omega t}$ obtained after a transitory given by:

$$i(t) = I(t)e^{\,j\omega t}$$

$$I(t) : 0 \rightarrow I_0$$

The current $I(t)$ goes from 0 to I_0 at the beginning. The impedance during the transitory time is:

$$Z(p) = Z_0 + \frac{\partial Z_0}{\partial p} \Delta p$$

Remember that:

$$p = \alpha + j\omega$$

and differentiation gives:

$$dp = d\alpha + j d\omega$$

whereas a variation gives:

$$\Delta p = \Delta\alpha + j\Delta\omega$$

This is:

$$\Delta p = (\alpha - \alpha_0) + j(\omega - \omega_0)$$

The circuit is linear when:

– We have no variation of frequency: $\omega = \omega_0$.

Suppose that losses are zero at the beginning $\alpha_0 = 0$.

Then

$$p = j\omega$$

and

$$\Delta p = \alpha$$

which gives us the result:

$$\Delta p = \alpha = \frac{1}{I(t)} \frac{d\, I(t)}{dt}$$

Then

$$Z(p) = Z_0 - j \frac{\partial Z_0}{\partial \omega} \frac{1}{I(t)} \frac{d\, I(t)}{dt}$$

and the voltage at the bounds of the linear impedance is written as:

$$V(t)\, e^{j\omega_0 t} = I(t) \left[\overbrace{Z_0 - j \frac{\partial Z_0}{\partial \omega} \frac{1}{I(t)} \frac{d\, I(t)}{dt}}^{Z(p)} \right] e^{j\omega_0 t}$$

The real power is expressed as:

$$P = \frac{1}{2} R_e \left\{ V^*(t)\, I(t) \right\}$$

This is:

$$P = \frac{1}{2} R_e \left\{ Z_0^* + j \frac{\partial Z_0^*}{\partial \omega} \frac{1}{I(t)} \frac{dI(t)}{dt} \right\} I^2(t)$$

and with a complex load $Z_0 = R_0 + j\, X_0$:

$$P = \frac{1}{2} \left[R_0 + \frac{\partial X_0}{\partial \omega} \frac{1}{I(t)} \frac{dI(t)}{dt} \right] I^2(t)$$

It can be written as:

$$P\,dt = \frac{1}{2}\,R_0\,I^2(t)\,dt \ + \frac{1}{2}\frac{\partial X_0}{\partial \omega}\,I(t)\,d\,I(t)$$

The first term is the dissipated energy in the resistance of the linear impedance, while the second term is the accumulate energy in the circuit.

If $I(t)$ goes from 0 to I_0, the circuit stores total energy:

$$E_{em} = \frac{1}{2}\frac{\partial X_0}{\partial \omega}\int_0^{I_0}I(t)\,dI(t) = \frac{1}{4}\frac{\partial X_0}{\partial \omega}\,I_0^2$$

The average energy dissipated by a cycle in the resistance is:

$$E_{dissipated} = \frac{1}{2}\,R_0\,I_0^2\,T_0 = \frac{1}{2}\frac{R_0\,I_0^2}{f_0}$$

Now we define the overvoltage of the linear circuit as:

$$Q = 2\pi\,\frac{E_{em}}{E_{dissipated}}$$

It results in:

$$Q = 2\pi\,\frac{\dfrac{1}{4}\dfrac{\partial X_0}{\partial \omega}\,I_0^2}{\dfrac{1}{2}\dfrac{R_0}{f_0}\,I_0^2}$$

or

$$Q = \frac{2\pi\,f_0}{2\,R_0}\frac{\partial X_0}{\partial \omega}\bigg|_{\omega=\omega_0}$$

which is:

$$Q = \frac{\omega_0}{2 R_0} \frac{\partial X_0}{\partial \omega} \Big|_{\omega = \omega_0}$$

Now if the circuit is formed by a loss resistance R_p and a useful resistance R_u, then:

– we define the empty overvoltage coefficient as:

$$Q_{empty} = 2\pi \frac{E_{em}}{E_{dissipated} \text{ in } R_p \text{ by cycle}} = \frac{\omega_0}{2 R_p} \frac{\partial X_0}{\partial \omega} \Big|_{\omega = \omega_0}$$

– the external overvoltage coefficient is written as:

$$Q_{ext} = 2\pi \frac{E_{em}}{E_{dissipated} \text{ in } R_u \text{ by cycle}} = \frac{\omega_0}{2 R_u} \frac{\partial X_0}{\partial \omega} \Big|_{\omega = \omega_0}$$

– the load overvoltage coefficient is written as:

$$Q_L = 2\pi \frac{E_{em}}{E_{dissipated} \text{ total by cycle}} = \frac{\omega_0}{2 (R_u + R_p)} \frac{\partial X_0}{\partial \omega} \Big|_{\omega = \omega_0}$$

With obviously:

$$\frac{1}{Q_L} = \frac{1}{Q_{ext}} + \frac{1}{Q_{empty}}$$

13.6.2. Overvoltage coefficients of an NL circuit

In this case, an oscillator is composed of an active element with a real negative part Z_{NL}, a utilization impedance R_u and a loss resistance R_p.

The total stored energy cannot be written:

$$E_{em} \neq \frac{1}{4} \frac{\partial X_0}{\partial \omega} I_0^2$$

But we continue to write by convention:

$$Q_{empty} = \frac{\omega_0}{2 R_p} \frac{\partial X_T}{\partial \omega}\bigg|_{\omega_0}^{I_0}$$

$$Q_{ext} = \frac{\omega_0}{2 R_u} \frac{\partial X_T}{\partial \omega}\bigg|_{\omega_0}^{I_0}$$

when we consider a current with an amplitude I_0 and a frequency ω_0.

With:

$$X_T = I_m (Z_T)$$
$$Z_T = Z_{NL} + R_P + R_u$$

and keeping in mind that for an oscillator, we have:

$$\sum R = 0$$

which gives:

$$|R_{NL}| = R_u + R_P$$

13.7. Bibliography

[BAH 03] BAHL I.J., BARTHIA P., *Microwave Solid State Circuit Design*, Wiley, 2003.

[CHA 94] CHANG K., *Microwave Solid-State Circuits and Applications*, Wiley-Interscience, 1994.

[GEN 84] GENTILI C., *Amplificateurs et Oscillateurs Microondes*, Masson, 1984.

[GRE 07] GREBENNIKOV A., *RF and Microwave Transistor Oscillator Design*, Wiley, 2007.

[JAR 04] JARRY P., Circuits Actifs Microondes: Amplificateurs, Oscillateurs, University of Bordeaux, 2004.

[PEN 88] PENNOCK S.R., SHEPHERD P.R., *Microwave Engineering with Wireless Applications*, McGraw-Hill Telecommunications, 1988.

[SOA 88] SOARES R., *GaAs MESFET Circuit Design*, Artech House, 1988.

Synchronization, Pulling and Spectra

14.1. Introduction

In this chapter, we show how to synchronize an oscillator, compute the variation in the frequency of the oscillation by varying the load (the pulling) and give the equivalent sinusoids in the cases of frequency-modulated (FM) and amplitude-modulated (AM) noises by considering the different spectrum.

14.2. Synchronization

In the microwave domain, it is possible to synchronize an oscillator by using another stable but low-power oscillator with a circulator.

We use a circulator to avoid returns on the synchronization generator.

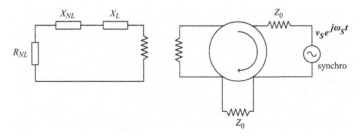

Figure 14.1. *The synchronization of a microwave oscillator*

The equivalent circuit is now given by the simple scheme (Figure 14.2).

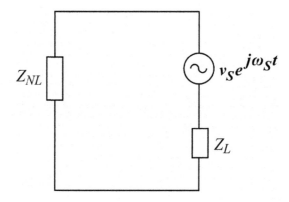

Figure 14.2. *Equivalent circuit of the synchronized oscillator*

The total impedance is $Z_T = Z_{NL} + Z_L$.

The evolution equation of the oscillators is also written as:

$$\left[Z_T \left(j\omega_0, I_0 \right) + \frac{\partial Z_T}{\partial p} dp + \frac{\partial Z_T}{\partial I} dI \right] I_0\, e^{j\omega_s t} = 0$$

where the oscillator is designed by a current, which goes from I_0 to $I_0 + \Delta I$:

$$I_0 \Rightarrow I_0 + \Delta I$$

The variation in the frequency can be important, and then:

$$dp \Rightarrow \Delta p$$

and the frequency (or phase) is modified as:

$$\omega_0 t \Rightarrow \omega_0 t + \varphi(t)$$

where ω_S is the synchronization frequency and Δp and ΔI are the variations of the complex frequency and intensity of the current.

$$\underbrace{\left[Z_T \left(j\omega_0, I_0 \right) + \frac{\partial Z_T}{\partial p} \Delta p + \frac{\partial Z_T}{\partial I} \Delta I \right] \left(I_0 + \Delta I \right) e^{(j\omega_0 t + \varphi lt)}}_{\text{oscillator}} = \underbrace{v_S \, e^{j\omega_s t}}_{\text{generator of synchronization}}$$

But from Chapter 13, the complex frequency is:

$$p = \alpha + j\omega = \frac{1}{I_0(t)} \frac{dI}{dt} + j \frac{d\varphi}{dt}$$

and the variation in this complex frequency is written as:

$$\Delta p = \Delta\alpha + j\Delta\omega = \frac{1}{I_0(t)} \frac{d\Delta I(t)}{dt} + j \frac{d\Delta\varphi}{dt}$$

And the first equation can be written with the operator $p = j\omega$ and $\dfrac{\partial}{\partial p} = - j \dfrac{\partial}{\partial \omega}$:

$$\left\{ Z_T \left(j\omega_0, I_0 \right) - j \frac{\partial Z_T}{\partial \omega} \left(\frac{1}{I_0} \frac{d\Delta I}{dt} + j \frac{d\Delta\Phi}{dt} \right) + \frac{\partial Z_T}{\partial I} \Delta I \right\}$$
$$\left(I_0 + \Delta I \right) e^{j(\omega_0 t + \Phi(t))} = v_S \, e^{j\omega_s t}$$

But we have still oscillations, then $Z_T \left(j\omega_0, I_0 \right) = 0$.

We neglect ΔI before I_0: $\Delta I \ll I_0$.

Let us write $\Omega = \omega_0 - \omega_S$, where ω_S is the frequency of the synchronization.

We say that the frequency of the output ω_0 gets to the frequency of the synchronization ω_S. The distance between them is Ω.

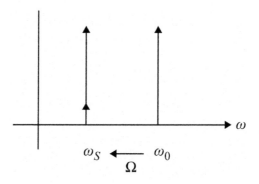

Figure 14.3. ω_0 *goes on* ω_S

Using $Z_T = R_T + jX_T$, we separate the real and imaginary parts of the formula, and after computations, we get:

$$\begin{cases} \dfrac{1}{I_0}\dfrac{\partial X_T}{\partial \omega}\dfrac{d\Delta I}{dt} + \dfrac{\partial R_T}{\partial \omega}\dfrac{d\Delta\varphi}{dt} + \dfrac{\partial R_T}{\partial I}\Delta I = \dfrac{v_s}{I_0}\cos(\Omega t + \varphi) \\[3mm] \dfrac{1}{I_0}\dfrac{\partial R_T}{\partial \omega}\dfrac{d\Delta I}{dt} - \dfrac{\partial X_T}{\partial \omega}\dfrac{d\Delta\varphi}{dt} - \dfrac{\partial R_T}{\partial I}\Delta I = \dfrac{v_s}{I_0}\sin(\Omega t + \varphi) \end{cases}$$

The state is stable if $\Delta I = Ct$, then $\dfrac{d\,\Delta I}{dt} = 0$ and also:

$$\varphi_0 = ct = \Omega t + \varphi(t)$$

which means:

$$\frac{d\varphi_0}{dt} = \Omega + \frac{d\varphi}{dt} = \Omega + \frac{d\Delta\varphi}{dt} = 0$$

Then

$$\frac{d\,\Delta\varphi}{dt} = -\Omega$$

Then, it becomes:

$$\begin{cases} -\Omega \dfrac{\partial R_T}{\partial \omega} + \dfrac{\partial R_T}{\partial I}\, \Delta I = \dfrac{v_s}{I_0}\cos \varphi_0 \\[2mm] \Omega \dfrac{\partial X_T}{\partial \omega} - \dfrac{\partial X_T}{\partial I}\, \Delta I = \dfrac{v_s}{I_0}\sin \varphi_0 \end{cases}$$

From the above two equations, we can obtain ω as a function of ΔI by eliminating $\cos \varphi_0$ and $\sin \varphi_0$ (with $\cos^2 \varphi_0 + \sin^2 \varphi_0 = 1$). We obtain the equation of an ellipse into two parts: a stable part and an unstable part (Figure 14.4).

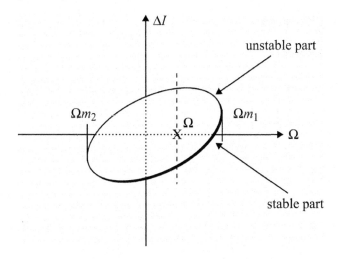

Figure 14.4. *Stable and unstable parts*

The current oscillator has the same frequency as the synchronization ω_s:

$$\begin{cases} i(t) = (I_0 + \Delta I)\, e^{j(\omega_0 t + \varphi)} \\[2mm] \text{with } \dfrac{d\varphi}{dt} = -\Omega \text{ then } \varphi = -\Omega t \end{cases}$$

which means that:

$$i(t) = (I_0 + \Delta I)e^{j\omega_S t}$$

The oscillator ω_0 is drive on the band of synchronization Ω by the generator of synchronization ω_S.

We can also determine the limits of the synchronization Ω_{m1} and Ω_{m2} and also the synchronization band $\Delta\omega = \Omega_{m1} - \Omega_{m2}$, which corresponds to a variation maximum of the current ΔI.

14.3. Pulling factor

14.3.1. *Definition*

The pulling factor characterizes the frequency variation in the oscillation when the load is varying. This variation is also called the involved frequency factor.

Let us consider (Figure 14.5) a stable oscillator loaded by the impedance Z_L.

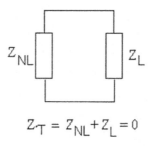

$$Z_T = Z_{NL} + Z_L = 0$$

Figure 14.5. *A stable oscillator*

If we vary the load impedance:

$$\Delta Z_L = \Delta R_L + j\Delta X_L$$

and the oscillator is still working, then we have:

$$\Delta Z_L + \frac{\partial Z_T}{\partial \omega} \Delta \omega + \frac{\partial Z_T}{\partial I} \Delta I = 0$$

with a current variation ΔI and a frequency variation $\Delta \omega$ where $\omega = \omega_0 \pm \Delta \omega$.

We separate the real and imaginary parts as:

$$\begin{cases} \Delta R_L + \dfrac{\partial R_T}{\partial \omega} \Delta \omega + \dfrac{\partial R_T}{\partial I} \Delta I = 0 \\[3mm] \Delta X_L + \dfrac{\partial X_T}{\partial \omega} \Delta \omega + \dfrac{\partial X_T}{\partial I} \Delta I = 0 \end{cases}$$

which give the frequency variation $\Delta \omega$ by eliminating the variation in the current ΔI :

$$\Delta \omega = \frac{\Delta R_L \dfrac{\partial X_T}{\partial I} - \Delta X_L \dfrac{\partial R_T}{\partial I}}{\dfrac{\partial R_T}{\partial I} \dfrac{\partial X_T}{\partial \omega} - \dfrac{\partial R_T}{\partial \omega} \dfrac{\partial X_T}{\partial I}}$$

This is the "pulling" factor that is the maximum relative deviation of frequency $\dfrac{\Delta \omega_r}{\omega_0}$ and corresponds to a load variation $\Delta Z_L = \Delta R_L + j \Delta X_L$.

14.3.2. Load variation

We use the following circuit (Figure 14.6) where R_0 is the characteristic of a line closed on $R_0 + \Delta R_0$ at the electric distance θ.

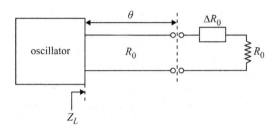

Figure 14.6. *Variation in the load*

Then, the load is:

$$Z_L = R_0 \frac{(R_0 + \Delta R_0) + j R_0 \, tg \, \theta}{R_0 + j (R_0 + \Delta R_0) tg\theta}$$

We can note that if $\Delta R_0 = 0$, then $Z_L = Z_0$ for all the electric length θ.

The return losses are characterized by a Voltage Standing Waves Ratio (VSWR) S:

$$S = \frac{R_0 + \Delta R_0}{R_0} \qquad S = 1 + \frac{\Delta R}{R_0}$$

and the frequency variation is given by:

$$\Delta\omega = B (S - 1) \frac{-\alpha S \, tg^2\theta + (S + 1) tg\theta + \alpha}{1 + S^2 \, tg \, \theta}$$

where the different constants are:

$$B = \frac{\dfrac{\partial R_T}{\partial I} R_0}{\dfrac{\partial R_T}{\partial I} \dfrac{\partial X_T}{\partial \omega} - \dfrac{\partial R_T}{\partial \omega} \dfrac{\partial X_T}{\partial I}}$$

and

$$\alpha = \frac{\partial X_T / \partial I}{\partial R_T / \partial I}$$

If $\dfrac{\partial R_T}{\partial \omega} = 0$, the total resistance is independent of the frequency and

if $\dfrac{\partial X_T}{\partial I} = 0$, the susceptance is also independent of the current and we have:

$$\Delta \omega_{\text{total}} = \frac{\omega_0}{2 Q_{\text{ext}}} \frac{S^2 - 1}{S}$$

Then, it is easy to find the frequency variation when the load is varying, and we have the "pulling" of the oscillator.

14.4. The spectrum of the oscillator

A spectrum of a perfect oscillator (without noise) is shown in Figure 14.7. There only one infinite beam at the frequency F_0.

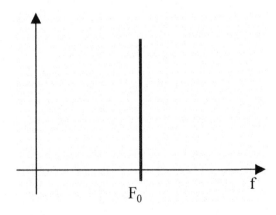

Figure 14.7. *Spectrum of a perfect oscillator*

But we have a real oscillator and the oscillator is noisy and its power spectrum is of the form as shown in Figure 14.8.

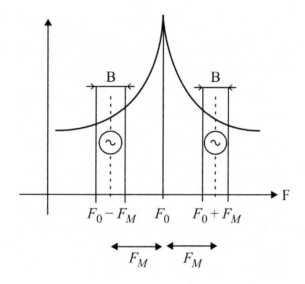

Figure 14.8. *Spectrum of a real (noisy) oscillator*

This is the spectrum of an AM wave and an FM wave. These two modulations are noise modulations.

Let us divide the "continuum" of the frequencies in a narrow band of width B (B is called the width of the window).

The two modulations (amplitude and frequency) are weak. The indices modulations are also weak. The power noise for all bands B can be given by a sinus equivalent signal with the same noise.

The noise power for all the bands B is then given by the signal:

$$V(t) = V_C \overbrace{\left[1 + m \cos\left(2\pi F_M\, t + \varphi_M\right) \right]}^{\text{AM}} . \cos \overbrace{\left[2\pi F_0 t + \beta \sin\left(2\pi F_M t + \phi_M\right) \right]}^{\text{FM}}$$

where m and β are, respectively, the amplitude modulation and frequency modulation. There is infinity of B bands.

14.4.1. Frequency modulation noise

Let us consider the phase modulation spectrum by noise. It has a density spectrum:

$$G_\varphi(F_M)$$

and the power in the B short band centered at F_M of the carrier F_0:

$$P = G_\varphi(F_M) \cdot B$$

But the power of the equivalent sinusoid with a low frequency modulation β is written as:

$$P = \frac{\beta^2}{4}$$

and we define the noise FM equivalent sinusoid by:

$$\frac{\beta^2}{4} = G\varphi(F_M) \cdot B$$

This gives:

$$\frac{(\Delta f)^2}{4 F_M^2} = G\varphi(F_M) \cdot B$$

where $\Delta f = \beta F_M$ is the maximum deviation of the FM:

$$\Delta f = 2 \sqrt{G\varphi(F_M)} \cdot \sqrt{B} \cdot F_M$$

and the efficacy deviation is given by:

$$\Delta f_{\text{eff}} = \frac{\Delta f}{\sqrt{2}} = \sqrt{2\,G\varphi\left(F_M\right)} \cdot \sqrt{B} \cdot F_M$$

14.4.2. Amplitude modulation noise

In the case of the amplitude modulation, we have a density spectrum $G_{\text{AM}}\left(F_M\right)$ and a power $P = G_{\text{AM}}\left(F_M\right) \cdot B$. We define the FM noise equivalent sinusoid by matching the power:

$$\frac{m^2}{4} = G_{\text{AM}}\left(F_M\right) \cdot B$$

and we can obtain the noise AM equivalent sinusoid.

14.5. Bibliography

[BAH 03] BAHL I.J., BARTHIA P., *Microwave Solid State Circuit Design*, Wiley, 2003.

[CHA 94] CHANG K., *Microwave Solid-State Circuits and Applications*, Wiley-Interscience, 1994.

[GEN 84] GENTILI C., *Amplificateurs et Oscillateurs Microondes*, Masson, 1984.

[GRE 07] GREBENNIKOV A., *RF and Microwave Transistor Oscillator Design*, Wiley, 2007.

[JAR 90] JARRY P., Microwave Oscillators, University of Brest and ENSTBr, 1990.

[JAR 04] JARRY P., Circuits Actifs Microondes: Amplificateurs, Oscillateurs, University of Bordeaux, 2004.

[PEN 88] Pennock S.R., Shepherd P.R., *Microwave Engineering with Wireless Applications*, McGraw-Hill Telecommunications, 1988.

[SOA 88] Soares R., *GaAs MESFET Circuit Design*, Artech House, 1988.

Integrated and Stable Microwave Oscillators Using Dielectric Resonators and Transistors

15.1. Introduction

A dielectric resonator (DR) is a ceramic microwave resonator with equivalent characteristics as a microwave cavity but with weak dimensions. $TE_{01\delta}$ is the fundamental mode and then the DR is equivalent to a magnetic dipole (Figure 15.1).

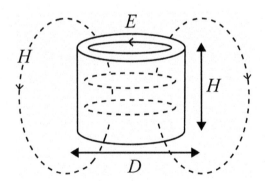

Figure 15.1. *The DR as a magnetic dipole*

The dimensions are about $\lambda_g = \dfrac{\lambda_0}{\sqrt{\varepsilon_r}}$, where ε_r is the relative electric constant, which is important (about 40). Also the important quality factor of the DR alone Q_0 involves that the energy is confined in the resonator, and radiation losses are weak.

In general, the used material is the barium titanate ($Ba_2Ti_9O_{23}$) and the oscillations go from 2 to 20 GHz.

15.2. A DR coupled to a microstrip line

15.2.1. *The scattering matrix*

The coupling is a function of the distance d between the DR and line of coupling (Figure 15.2).

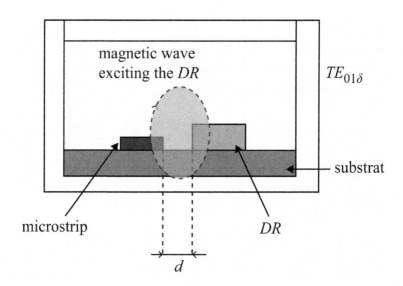

Figure 15.2. *Coupling a DR*

Then, the equivalent circuit of the DR coupled with a transmission line is shown in Figure 15.3.

In fact, the DR is a reflection cavity and it reflects all the microwave energy at the resonance. The coupling coefficient in the plane (P, P') is then given by:

$$\beta = \frac{R}{R_{\text{exterior}}} = \frac{R}{2Z_0}$$

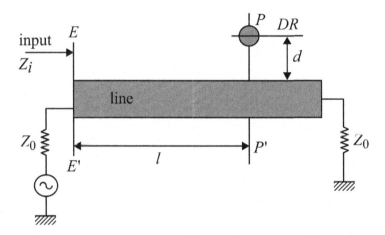

Figure 15.3. *DR coupled with a line*

When the quantity d grows, then the coupling β decreases and we show that the input normalized impedance is of the form:

$$z_i = \frac{Z_i}{Z_0} = \frac{Z_0 + Z}{Z_0} = 1 + \frac{Z}{Z_0} = 1 + \frac{2\beta}{1 + j2Q_{\text{empty}}\dfrac{\Delta\omega}{\omega_0}}$$

where Q_{empty} is the empty overvoltage coefficient, β is the coupling coefficient and $\Delta\omega = \omega - \omega_0$.

The localized equivalent circuit is given when considering only the plane PP' as shown in Figure 15.4.

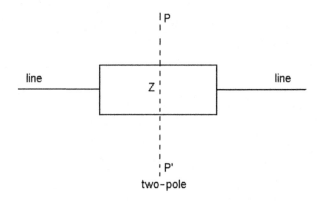

Figure 15.4. *Coupling with the line*

Then, we have in the plane PP' :

$$\begin{cases} S_{11} + S_{21} = 1 \\ S_{11} = \dfrac{z_e - 1}{z_e + 1} \end{cases}$$

Then

$$S_{11} = \frac{\beta}{1 + \beta + 2j\,Q_u\,\dfrac{\Delta\omega}{\omega_0}}$$

and

$$S_{21} = 1 - S_{11} = \frac{1 + 2j\,Q_u\,\dfrac{\Delta\omega}{\omega_0}}{1 + \beta + 2j\,Q_u\,\dfrac{\Delta\omega}{\omega_0}}$$

This gives the scattering matrix of the DR coupled to the line:

$$(S) = \frac{1}{1 + \beta + 2j\,Q_u\,\dfrac{\Delta\omega}{\omega_0}} \begin{pmatrix} \beta & \left| 1 + 2j\,Q_u\,\dfrac{\Delta\omega}{\omega_0} \right. \\ 1 + 2j\,Q_u\,\dfrac{\Delta\omega}{\omega_0} & \left. \beta \right. \end{pmatrix}$$

15.2.2. *The interpretation*

First, if we consider a circuit without coupling $\beta = 0$, there is no reflection $S_{11} = 0$, and we have a perfect transmission $S_{21} = 1$. This corresponds to the case where the distance d of the DR to the line is infinity (Figure 15.5).

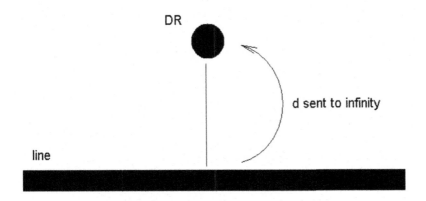

Figure 15.5. *The distance d is infinity*

Second, if we have a maximum coupling $\beta = \infty$, all the power is going through the DR and the scattering matrix is a unity matrix $(S) = \begin{pmatrix} 1 & 0 \\ 0 & 1 \end{pmatrix}$.

And if we are at the center frequency $\omega = \omega_0$, the reflection factor and the transmission factor are only a function of the coupling β :

$$S_{11} = \frac{\beta}{1+\beta} \text{ and } S_{21} = \frac{1}{1+\beta}$$

15.2.3. *The influence of the length l on the transmission line*

The length l of the transmission line induces an electric length $\theta = \beta' l$, where β' is the propagation constant. The two planes P, P' and E, E' are distant from l (Figure 15.6).

We have from the formulas from the input impendence is:

$$Z_e = Z_0 \frac{Z_L + j Z_0 \, tg \, \theta}{Z_0 + j Z_L \, tg \, \theta}$$

where the load Z_L is the impedance formed by Z and Z_0 in parallel:

$$Z_L = Z \,/\!/\, Z_0$$

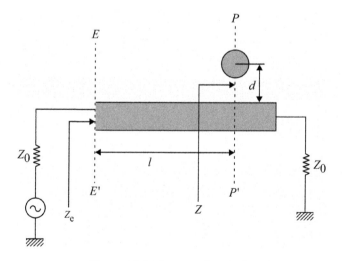

Figure 15.6. *The input plane E, E'*

Then, we see in the input plane E, E' :

$$Z_e = Z_0 \frac{1 + j(1 + Y Z_0) \, tg \, \theta}{(1 + Y Z_0) + j \, tg \, \theta}$$

where

$$Y = \frac{1 + 2jQ_u \dfrac{\Delta\omega}{\omega_0}}{2\,\beta\,Z_0}$$

θ, $\Delta\omega$ and β are variables that we can change to create an oscillator. We can also change the input impedance Z_e by varying θ (then the distance l). Then, we can obtain all kinds of input impedances Z_e (Figure 15.7).

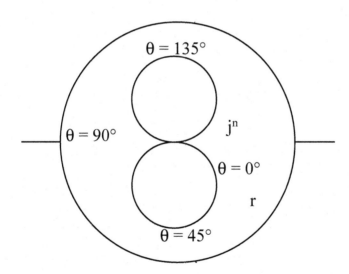

Figure 15.7. *The input plane E,E' in the Smith plane*

15.3. Realization of a stable oscillator with a DR

The DR has low losses, an important quality factor, good temperature stability and it can be miniaturized. It is a good component that allows us to stabilize in frequency of the field-effect transistor (FET) oscillators.

15.3.1. *Configuration*

We use the configuration as shown in Figure 15.8.

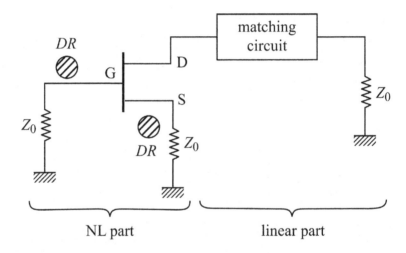

Figure 15.8. *The used configuration*

Using one or several DR, it is possible to match the impedance and stabilize the oscillator. With the matching circuit, we can fix the oscillation condition and have a maximum of power at the output.

Let us consider the transistor as a three-port and the different impedances to be match need an experiment and previous characterization of the transistor.

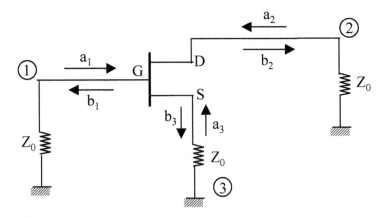

Figure 15.9. *The transistor alone*

15.3.2. *Characterization of the transistor*

The characterization of the FET is made with S parameters (Figure 15.9). The parameters S_{ij} are given by measure:

$$\begin{cases} b_1 = S_{11}\, a_1 + S_{12}\, a_2 + S_{13}\, a_3 \\ b_2 = S_{21}\, a_1 + S_{22}\, a_2 + S_{23}\, a_3 \\ b_3 = S_{31}\, a_1 + S_{32}\, a_2 + S_{33}\, a_3 \end{cases}$$

and here we also have:

$$\sum_{j=1}^{3} S_{ij} = \sum_{i=1}^{3} S_{ij} = 1$$

15.3.3. *Determination of the source impedance Z_3*

We determine the source impedance Z_3, which gives a maximum reflection coefficient in the drain S_{22} (Figure 15.10).

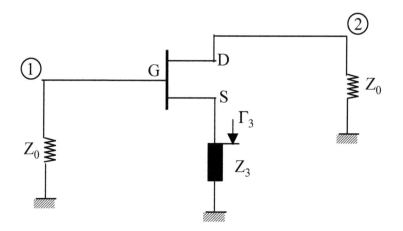

Figure 15.10. *Source impedance*

Knowing Z_3, the transistor component (three-port) is transformed into an instable two-port.

We have:

$$\Gamma_3 = \frac{a_3}{b_3} = \frac{Z_3 - Z_0}{Z_3 + Z_0}$$

Then

$$\begin{cases} b_1 = S_{11} \, a_1 + S_{12} \, a_2 + S_{13} \, b_3 \, \Gamma_3 \\ b_2 = S_{21} \, a_1 + S_{22} \, a_2 + S_{23} \, b_3 \, \Gamma_3 \\ b_3 = S_{31} \, a_1 + S_{32} \, a_2 + S_{33} \, b_3 \, \Gamma_3 \end{cases}$$

and

$$b_3 = \frac{S_{31} \, a_1 + S_{32} \, a_2}{1 - S_{33} \, \Gamma_3}$$

Then, we come to a 2×2 matrix:

$$\begin{cases} b_1 = S_{11}\, a_1 + S_{12}\, a_2 + \dfrac{S_{31}\, a_1 + S_{32}\, a_2}{1 - S_{33}\, \Gamma_3}\, S_{13}\, \Gamma_3 \\[4mm] b_2 = S_{21}\, a_1 + S_{22}\, a_2 + \dfrac{S_{31}\, a_1 + S_{32}\, a_2}{1 - S_{33}\, \Gamma_3}\, S_{23}\, \Gamma_3 \end{cases}$$

and

$$\begin{pmatrix} b_1 \\ b_2 \end{pmatrix} \begin{pmatrix} S_{11} + \dfrac{S_{13}\, S_{31}\, \Gamma_3}{1 - S_{33}\, \Gamma_3} & S_{12} + \dfrac{S_{13}\, S_{32}\, \Gamma_3}{1 - S_{33}\, \Gamma_3} \\[5mm] S_{21} + \dfrac{S_{23}\, S_{31}\, \Gamma_3}{1 - S_{33}\, \Gamma_3} & S_{22} + \dfrac{S_{23}\, S_{32}\, \Gamma_3}{1 - S_{33}\, \Gamma_3} \end{pmatrix} \begin{pmatrix} a_1 \\ a_2 \end{pmatrix}$$

where

$$(b) = \left(S^T \right)(a)$$

where $\left(S^T \right)$ is the scattering matrix of the two-port transistor loaded by the impedance Z_3. S_{ij}^T are the parameters of this two-port given as a function of the reflection coefficient Γ_3.

$$S_{ij}^T = Sij + \frac{S_{i3}\, S_{3j}\, \Gamma_3}{1 - S_{33}\ \Gamma_3}$$

When Γ_3 move on the whole Smith chart, the four coefficients $S_{11}^T, S_{12}^T, S_{21}^T$ and S_{22}^T take the values shown in Figure 15.11. From a simple reading, we can find the values of the parameters, which give the more instability.

We give an example: suppose we have at 9 GHz the next response (Figure 15.11).

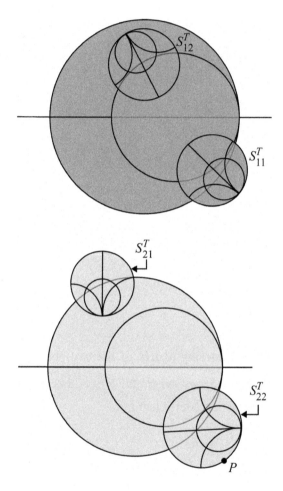

Figure 15.11. *Instability on the Smith chart*

In this example, the point P gives the more instability in vicinity of S_{22}^T. This means that $|S_{22}| > 1$ and $Z_0 \le 0$ (Z_0 is negative). In this choose example (point P) we have $\left|S_{22}^T\right|=1.4$ which corresponds to a capacitance of 0.25 pF and Z_3 is known (Figure 15.12).

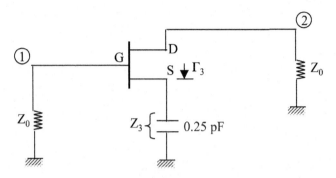

Figure 15.12. *Determination of Z_3*

To get more stability, Z_3 is realized by a DR (Figure 15.13).

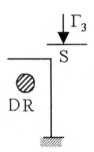

Figure 15.13. *Realization using a DR*

Then, the reflection coefficient Γ_3 is given by:

$$\Gamma_3 = \frac{\beta}{\sqrt{(\beta+1)^2+(2\,Q_u)^2}}\exp\left[-2j\left(\theta+Arctg\frac{2\,Q_u\,\delta}{(\beta+1)}\right)\right]$$

where the normalized frequency ratio is:

$$\delta = \frac{f - f_0}{f_0}\, v$$

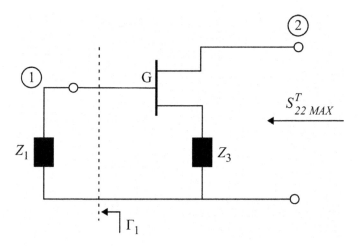

Figure 15.14. *Determination of Z_1*

15.3.4. *Determination of the gate impedance Z_1*

We know the S parameters of the two-port loaded by the impedance Z_3. Then, it is possible to determine the impedance Z_1, which gives the maximum of $S^T_{22\,MAX}$.

$$\begin{cases} S^T_{22\,MAX} = S^T_{11} + \dfrac{S^T_{12}\, S^T_{21}\, \Gamma_1}{1 - S^T_{22}\, \Gamma_1} \\[2ex] \Gamma_1 = \dfrac{z_1 - z_0}{z_1 + z_0} \end{cases}$$

We also have a graphical determination of Z_1 and this impedance can be also realized using a DR.

15.3.5. *Determination of the load impedance* Z_L

$S^T_{22\,MAX}$ is at its maximum, and the load Z_L is determined so that the oscillation condition is satisfied.

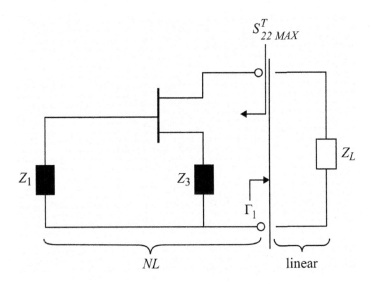

Figure 15.15. *Determination of* Z_L

And we must have:

$$\begin{cases} \left| S^T_{22\,MAX} \right| \cdot \left| \Gamma_L \right| \ge 1 \\ \angle S^T_{22\,MAX} + \angle \Gamma_L = 0 \end{cases}$$

The first equation (amplitude equation) is verified. The second equation (phase equation) has yet to be satisfied. R_L and X_L (with $Z_L = R_L + X_L$) are determined by measuring the output impedance of the system to have a given power.

Then, we trace all the outline power $-1\,\mathrm{db}, -2\,\mathrm{db}, -6\,\mathrm{db},$ etc. and Γ_L is taken on these curves in manner to realize

$\angle S_{22\,MAX}^{T} + \angle \Gamma_L = 0$, and we can get the maximum of the output power $P_0\left(MAX\right)$.

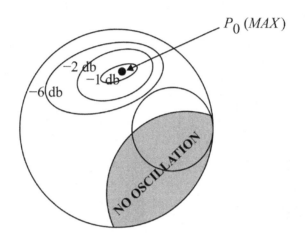

Figure 15.16. *The output power*

15.4. Bibliography

[BAH 03] BAHL I.J., BARTHIA P., *Microwave Solid State Circuit Design*, Wiley, 2003.

[CHA 94] CHANG K., *Microwave Solid-State Circuits and Applications*, Wiley-Interscience, 1994.

[GEN 84] GENTILI C., *Amplificateurs et Oscillateurs Microondes*, Masson, 1984.

[GRE 07] GREBENNIKOV A., *RF and Microwave Transistor Oscillator Design*, Wiley, 2007.

[JAR 90] JARRY P., Microwave Oscillators, University of Brest and ENSTBr, 1990.

[JAR 04] JARRY P., Circuits Actifs Microondes: Amplificateurs, Oscillateurs, University of Bordeaux, 2004.

[PEN 88] PENNOCK S.R., SHEPHERD P.R., *Microwave Engineering with Wireless Applications*, McGraw-Hill Telecommunications, 1988.

[SOA 88] SOARES R., *GaAs MESFET Circuit Design*, Artech House, 1988.

Problems

16.1. Scattering parameters of a transistor

The (S_{ij}) are the scattering parameters of a transistor (Figure 16.1).

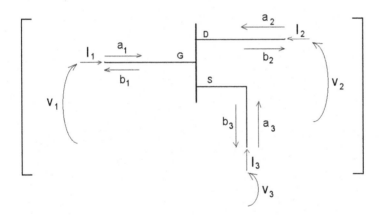

Figure 16.1. *The waves of a transistor*

1) Show that we have $\displaystyle\sum_{i=1}^{3} S_{ij} = 1, \quad j = 1, 2, 3.$

2) What happens in the case of a two-port?

The solution

1) By definition, the direct and reflected waves are expressed as:

$$
\begin{cases}
a_i = \dfrac{V_i + R_0 I_i}{2\sqrt{R_0}} \\[4mm]
b_i = \dfrac{V_i - R_0 I_i}{2\sqrt{R_0}}
\end{cases}
$$

where R_0 is the characteristic impedance. If the characteristic impedance is normalized $(R_0 = 1)$, we have:

$$
\begin{cases}
V_1 = a_1 + b_1 \\
V_2 = a_2 + b_2 \\
V_3 = a_3 + b_3
\end{cases}
\text{ and }
\begin{cases}
I_1 = a_1 - b_1 \\
I_2 = a_2 - b_2 \\
I_3 = a_3 - b_3
\end{cases}
$$

But $(b) = (S)(a)$, then:

$$
\begin{cases}
I_1 = a_1 - b_1 = a_1 - \left(S_{11}a_1 + S_{12}a_2 + S_{13}a_3\right) \\
I_2 = a_2 - b_2 = a_2 - \left(S_{21}a_1 + S_{22}a_2 + S_{23}a_3\right) \\
I_3 = a_3 - b_3 = a_3 - \left(S_{31}a_1 + S_{32}a_2 + S_{33}a_3\right)
\end{cases}
$$

and we have on the transistor:

$$
I_1 + I_2 + I_3 + 0
$$

which gives:

$$
a_1 + a_2 + a_3 = \left(S_{11} + S_{21} + S_{31}\right)a_1 + \left(S_{12} + S_{22} + S_{32}\right)a_2 + \left(S_{13} + S_{23} + S_{33}\right)a_3
$$

or

$$\begin{cases} S_{11} + S_{21} + S_{31} = 1 \\ S_{12} + S_{22} + S_{32} = 1 \\ S_{13} + S_{23} + S_{33} = 1 \end{cases}$$

i.e.

$$\sum_{i=1}^{3} S_{ij} = 1, \quad j = 1, 2, 3$$

2) In this case:

$$\begin{cases} b_1 = S_{11}a_1 + S_{12}a_2 \\ b_2 = S_{21}a_1 + S_{22}a_2 \end{cases}$$

we have $V_1 = V_2$, which gives:

$$a_1 + b_1 = a_2 + b_2$$

or

$$a_1 + \left(S_{11}a_1 + S_{12}a_2 \right) = a_2 + \left(S_{21}a_1 + S_{22}a_2 \right)$$

or

$$\begin{cases} 1 + S_{11} = S_{21} \\ 1 + S_{22} = S_{12} \end{cases}$$

If the circuit is reciprocal $S_{12} = S_{21}$, then:

$$S_{12} = S_{21} = 1 + S_{11} = 1 + S_{22}$$

16.2. Scattering parameters and oscillations conditions

1) *Scattering parameters for a two-parts*: Give the oscillation condition of a two-poles using its scattering parameters (S). We recall that an oscillator is characterized by a linear part of characteristic impedance Z_C (reflection coefficient γ_C) and by a nonlinear part Z_{NL} (reflection coefficient γ_{NL}). Z_0 is the characteristic impedance of the system.

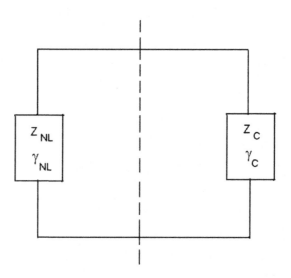

Figure 16.2. *Characterization of an oscillator*

2) We consider the modulus Γ and the phase $\angle\Gamma$ of the reflection coefficients $\gamma = \Gamma e^{\angle\Gamma}$. What is the condition with the two modulus Γ_{NL} and Γ_C and with the two phases $\angle\Gamma_{NL}$ and $\angle\Gamma_C$?

3) *Oscillations of a n–port*: Now we consider two *n*-ports. One is passive (linear) and the other is active (nonlinear). The incident waves are the a_i (or the a_i') and the reflected waves are the b_i (or the b_i').

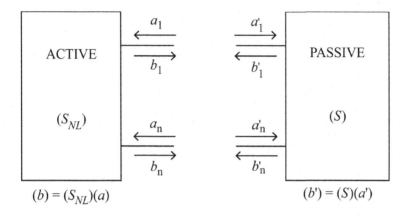

Figure 16.3. *Characterization of an n–port*

Connection of these two *n*-ports gives an oscillator. What are the (generalized) oscillation conditions?

4) Recover the case of the two-port.

5) *Oscillations of a four–port* An active four-port (nonlinear S_{NL}) is closed on two passive loads (linear and reflection coefficients Γ_1, Γ_2).

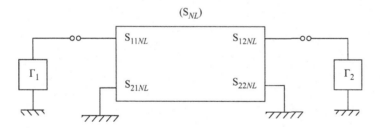

Figure 16.4. *Characterization of a four–port*

Give the scattering matrix (S) of the load (linear) using Γ_1 and Γ_2.

6) (S_{NL}) is known, then gives the condition of oscillation. This condition will be given as a function of $\Gamma_1, \Gamma_2, S_{11NL}, S_{12NL}, S_{21NL}$ and S_{22NL}.

The solution

1) *two-port*: From Figure 16.2, we must have:

$$Z_{NL} + Z_C = 0$$

But there is a relation between the impedances (Z_{NL}, Z_C) and the reflection coefficients (γ_{NL}, γ_C):

$$\begin{cases} Z_{NL} = \dfrac{1+\gamma_{NL}}{1-\gamma_{NL}} Z_0 \\ Z_C = \dfrac{1+\gamma_C}{1-\gamma_C} Z_0 \end{cases}$$

where Z_0 is the characteristic impedance of the system. This means that we must have:

$$\frac{1+\gamma_{NL}}{1-\gamma_{NL}} + \frac{1+\gamma_C}{1-\gamma_C} = 0$$

or

$$2\frac{1-\gamma_C \gamma_{NL}}{(1-\gamma_{NL})(1-\gamma_C)} = 0$$

It means that:

$$\gamma_C \gamma_{NL} = 1$$

2) But the complex numbers γ have a modulus and phase $\gamma = \Gamma e^{\angle \Gamma}$, then:

$$\begin{cases} \Gamma_{NL} + \Gamma_C = 1 & \text{mod ulus} \\ \angle \Gamma_{NL} + \angle \Gamma_C = 2k\pi & \text{phase} \end{cases}$$

In general, we take $k = 0$. This is the result for a two-port.

3) n–port

The two n-ports are connected to form an oscillator:

$$\begin{cases} (b') = (a) \\ (b) = (a') \end{cases}$$

Then

$$(b) = (S_{NL})(a)$$
$$(b) = (S_{NL})(b')$$
$$(b) = (S_{NL})(S)(a')$$

which gives:

$$(a') = (S_{NL})(S)(a')$$

or

$$\{(S_{NL})(S) - I_n\}(a') = 0$$

But the vector $(a') \neq 0$ and $(S_{NL})(S) - I_n$ are a singular matrix. Their determinant is zero:

$$\text{Det}\{(S_{NL})(S) - I_n\} = 0$$

This is the generalized condition of oscillation of an oscillator using an n-port active component.

4) In the case of two-port, the scattering matrices reduce to the reflection coefficients:

$$\begin{cases}(S_{NL}) \to \gamma_{NL} \\ (S) \to \gamma_C\end{cases}$$

and

$$\text{Det}\{\gamma_{NL}\gamma_C - 1\} = 0$$

or

$$\gamma_{NL} \cdot \gamma_C = 1$$

And we recover the two-port case of the first question.

5) *Four–port*

In this case, the linear scattering matrix (S) reduces to a 2×2 matrix:

$$(S) = \begin{pmatrix}\Gamma_1 & 0 \\ 0 & \Gamma_2\end{pmatrix}$$

6) With the nonlinear matrix (S_{NL}), we have:

$$(S_{NL})(S) - I_2 = \begin{pmatrix}\Gamma_1 S_{11NL} - 1 & \Gamma_2 S_{12} \\ \Gamma_1 S_{21} & \Gamma_2 S_{22NL} - 1\end{pmatrix}$$

And computing the determinant, we get the stability condition of the four-port.

$$\Gamma_1 \Gamma_2 \left(S_{11NL} S_{22NL} - S_{12NL} S_{21NL}\right) - \Gamma_1 S_{11NL} - \Gamma_2 S_{22NL} + 1 = 0$$

or

$$\Gamma_1\Gamma_2\Delta S_{NL} - \Gamma_1 S_{11NL} - \Gamma_2 S_{22NL} + 1 = 0$$

16.3. Synchronization of an oscillator

We synchronize a microwave power oscillator (ω_0, I_0) by a stable source with a low power oscillator (ω_S, V_S). This oscillator is drive ("entrainé") on a synchronization band Ω so that $\varphi_0 = cte$. The equation of evolution of the microwave oscillator is given by:

$$\begin{cases} -\Omega\dfrac{\partial R_T}{\partial \omega} + \dfrac{\partial R_T}{\partial I}\Delta I = \dfrac{V_S}{I_0}\cos\varphi_0 \\[2ex] \Omega\dfrac{\partial X_T}{\partial \omega} - \dfrac{\partial X_T}{\partial I}\Delta I = \dfrac{V_S}{I_0}\sin\varphi_0 \end{cases}$$

with

$$\varphi_0 = \Omega t + \varphi(t)$$

where ω_S is the frequency synchronization and the modified equivalent circuit of the oscillator is discussed in the following.

1) Give the band of synchronization Ω.

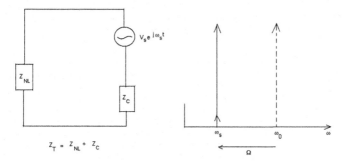

Figure 16.5. *Synchronization of an oscillator*

2) What are the limits of synchronization of the oscillator? And what is the maximum excursion of the frequency $\Delta\omega_r$?

3) What does this excursion become when $\partial X_T / \partial I = 0$? Give this expression as a function of the characteristics of the circuit:

$$\left\{ \begin{array}{l} Q_{ext} = \dfrac{\omega_0}{2R_C}\dfrac{\partial X_T}{\partial\omega} \\[2mm] P_{oscill} = \dfrac{1}{2}R_C I_0^2 \\[2mm] P_{synch} = \dfrac{1}{8}\dfrac{V_S^2}{R_C} \end{array} \right.$$

4) Determine the synchronization curve when $d = b = 1$ and $c = e = 3$.

5) What happens to this synchronization curve when $b = e = 1$ and $c = -d = 2$?

The solution

1) The band of synchronization Ω.

Let us consider:

$$a = \dfrac{V_S}{I_0}; \quad b = \dfrac{\partial X_T}{\partial I}; \quad c = \dfrac{\partial R_T}{\partial I}; \quad d = \dfrac{\partial R_T}{\partial\omega}; \quad e = \dfrac{\partial X_T}{\partial\omega}$$

Then, we have:

$$\left\{ \begin{array}{l} -d\,\Omega + c\,\Delta I = a\cos\varphi_0 \\ e\,\Omega - b\,\Delta I = a\sin\varphi_0 \end{array} \right.$$

We have two equations with two unknowns and it is possible to give the expression of Ω as a function of φ_0.

$$\Omega = a \frac{b\cos\varphi_0 + \sin\varphi_0}{ce - db}$$

i.e.

$$\Omega = \frac{V_S}{I_0} \frac{\dfrac{\partial X_T}{\partial I}\cos\varphi_0 + \dfrac{\partial R_T}{\partial I}\sin\varphi_0}{\dfrac{\partial R_T}{\partial I}\dfrac{\partial X_T}{\partial \omega} - \dfrac{\partial R_T}{\partial \omega}\dfrac{\partial X_T}{\partial I}}$$

2) Now we give the limits of synchronization. In fact, we have to determine the maximum(s) of Ω by computing the derivative $\partial\Omega/\partial\varphi_0$.

Let us consider:

$$\begin{cases} \alpha = \dfrac{\partial X_T/\partial I}{\partial R_T/\partial I} \\[3mm] K = \left(\dfrac{\partial R_T}{\partial I}\dfrac{\partial X_T}{\partial \omega} - \dfrac{\partial R_T}{\partial \omega}\dfrac{\partial X_T}{\partial I} \right) \end{cases}$$

Then

$$\Omega = \frac{a}{K}\frac{\partial R_T}{\partial I}\left[\alpha\cos\varphi_0 + \sin\varphi_0\right]$$

and

$$\frac{d\Omega}{d\varphi_0} = \frac{a}{K}\frac{\partial R_T}{\partial I}\left[-\alpha\sin\varphi_0 + \cos\varphi_0\right] = 0$$

The solution(s) is expressed as:

$$tg\varphi_{0\,MAX} = \frac{1}{\alpha} = \frac{\partial R_T/\partial I}{\partial X_T/\partial I}$$

In fact, there are two solutions that give this value using $\sin\varphi = \dfrac{\pm tg\varphi}{\sqrt{1+tg^2\varphi}}$ and $\cos\varphi = \dfrac{\pm 1}{\sqrt{1+tg^2\varphi}}$:

$$\begin{cases} \sin\varphi_{0\,MAX1} = \dfrac{1}{\sqrt{1+\alpha^2}}; \quad \text{with } \cos\varphi_{0\,MAX1} = \dfrac{\alpha}{\sqrt{1+\alpha^2}} \\ \text{and} \\ \sin\varphi_{0\,MAX2} = -\dfrac{1}{\sqrt{1+\alpha^2}}; \quad \text{with } \cos\varphi_{0\,MAX2} = -\dfrac{\alpha}{\sqrt{1+\alpha^2}} \end{cases}$$

which correspond to two maximums of Ω:

$$\begin{cases} \Omega_{MAX1} = \dfrac{a}{K}\dfrac{\partial R_T}{\partial I}\sqrt{1+\alpha^2} \\ \Omega_{MAX2} = -\dfrac{a}{K}\dfrac{\partial R_T}{\partial I}\sqrt{1+\alpha^2} \end{cases}$$

and the maximum excursion of the frequency $\Delta\omega_T$ is written as:

$$\Delta\omega_T = \Omega_{MAX1} - \Omega_{MAX2} = \frac{2\,a}{K}\frac{\partial R_T}{\partial I}\sqrt{1+\alpha^2}$$

3) $\alpha = 0$ and $K = \dfrac{\partial R_T}{\partial I}\dfrac{\partial X_T}{\partial\omega}$ when $\partial X_T/\partial I = 0$ and this excursion becomes:

$$\Delta\omega_T = 2\frac{V_S}{I_0}\frac{1}{\partial X_T/\partial\omega}$$

To find $\Delta\omega_T$ as a function of the characteristics of the circuit, we determine:

$$\sqrt{\frac{P_{\text{synch}}}{P_{\text{oscill}}}} = \frac{1}{2R_C}\frac{V_S}{I_0}$$

and

$$\Delta\omega_T = \frac{2\omega_0}{Q_{\text{ext}}}\sqrt{\frac{P_{\text{synch}}}{P_{\text{oscill}}}}$$

This excursion $\Delta\omega_T$ is proportional to the frequency ω_0 and $\sqrt{P_{\text{synch}}}$ and inverse to Q_{ext} and $\sqrt{P_{\text{oscill}}}$.

4) The synchronization curve.

Remember, we had:

$$\begin{cases} -d\,\Omega + c\,\Delta I = a\cos\varphi_0 \\ e\,\Omega - b\,\Delta I = a\sin\varphi_0 \end{cases}$$

Using the property $\cos^2\varphi_0 + \sin^2\varphi_0 = 1$, we make the term in φ_0 disappear:

$$\Omega^2\left(d^2 + e^2\right) - 2\Omega\left(\Delta I\right)\left(dc + eb\right) + \left(\Delta I\right)^2\left(b^2 + c^2\right) = a^2$$

This is the equation of a bent ellipse (Figure 16.6). In fact, we only have half an ellipse of stability because to a single frequency Ω corresponds only one current ΔI.

The bent axes (α,β) are deduced from the $(\Omega,\Delta I)$ from the following equations:

$$\begin{cases} \Omega = \alpha x - \beta y \\ \Delta I = \beta x + \alpha y \end{cases}$$

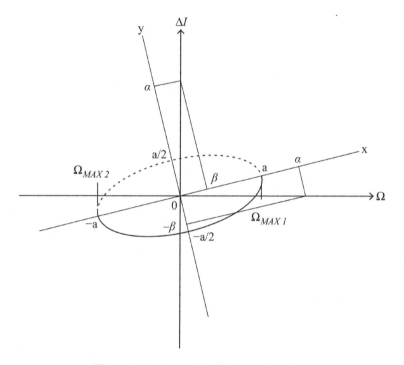

Figure 16.6. *Bent and half ellipse of stability*

This corresponds to an axis' rotation with α and $\beta \geq 0$.

To simplify the problem, let us take the particular case:

$$b = 1; c = 3; d = 1; e = 3$$

This gives a very simple equation of the bent ellipse:

$$5\Omega^2 - 6\Omega\left(\Delta I\right) + 5\left(\Delta I\right)^2 = \frac{a^2}{2}$$

and with the transformation:

$$\left(5\alpha^2 - 6\alpha\beta + 5\beta^2\right)x^2 - 6\left(\alpha^2 - \beta^2\right)xy + \left(5\alpha^2 + 6\alpha\beta + 5\beta^2\right)y^2 = 0$$

If we take $\alpha = \beta$ and $\alpha \geq 0$, $\beta \geq 0$, we have the equation of an unbent ellipse. To simplify the problem, let us also write:

$$\alpha = \beta = \frac{1}{2\sqrt{2}}$$

Then, we have:

$$\frac{x^2}{a^2} + \frac{y^2}{a^2/4} = 1$$

We recover the equation of a direct ellipse:

– with a small axis a ;

– with a big axis $b = a/2$.

In this case, the big axis is smaller than the small axis.

5) When $b = e = 1, c = -d = 2$ and taking $\alpha = 1/4$ after computation, we find that the synchronization curve becomes:

$$\frac{x^2}{a^2} + \frac{y^2}{4a^2} = 1$$

In this case, the small and big axes of the ellipse are, respectively, a and $2a$.

16.4. Pulling factor of an oscillator

The pulling factor is the frequency variation of the oscillation when the load is varying. Let us consider a stable oscillator loaded by the impedance Z_L (Figure 16.7).

The load impedance is varying:

$$\Delta Z_L = \Delta R_L + j \Delta X_L$$

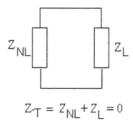

$$Z_T = Z_{NL} + Z_L = 0$$

Figure 16.7. *An oscillator with a load* Z_L

1) The oscillator is still working, what is the relation between ΔZ_L, $\Delta \omega$ and ΔI ?

2) From this relation, deduce the frequency variation.

3) To make a variation of the load, we use at the output of the oscillator (Figure 16.8) a line closed on $R_0 + \Delta R_0$ at the electric distance θ. R_0 is the characteristic impedance of the line. What is the value of the impedance of the load Z_L as a function of R_0, θ and the Voltage Standing Waves Ratio (VSWR) S?

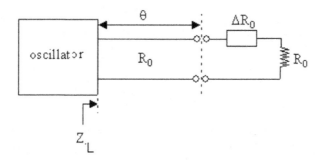

Figure 16.8. *Variation of the load*

4) What is the frequency variation $\Delta \omega$?

5) For what values of θ and S, is this frequency variation maximum ($\Delta\omega_1$ and $\Delta\omega_2$)? Then what is the maximum excursion of frequency $\Delta\omega_2 - \Delta\omega_1$.

6) What is the sensibility of the oscillator $(\Delta f_0 / f_0)$?

The solution

1) The oscillator is still working:

$$\Delta Z_L + \frac{\partial Z_T}{\partial \omega} \Delta\omega + \frac{\partial Z_T}{\partial I} \Delta I = 0$$

2) We separate the real and imaginary parts:

$$\begin{cases} \Delta R_L + \dfrac{\partial R_T}{\partial \omega} \Delta\omega + \dfrac{\partial R_T}{\partial I} \Delta I = 0 \\[2mm] \Delta X_L + \dfrac{\partial X_T}{\partial \omega} \Delta\omega + \dfrac{\partial X_T}{\partial I} \Delta I = 0 \end{cases}$$

which gives the frequency variation $\Delta\omega$ by eliminating the variation of the current ΔI :

$$\Delta\omega = \frac{\Delta R_L \dfrac{\partial X_T}{\partial I} - \Delta X_L \dfrac{\partial R_T}{\partial I}}{\dfrac{\partial R_T}{\partial I} \dfrac{\partial X_T}{\partial \omega} - \dfrac{\partial R_T}{\partial \omega} \dfrac{\partial X_T}{\partial I}}$$

3) Then, the load is:

$$Z_L = R_0 \frac{(R_0 + \Delta R_0) + jR_0 \, tg \, \theta}{R_0 + j(R_0 + \Delta R_0) \, tg\theta}$$

We can note that if $\Delta R_0 = 0$, then $Z_L = Z_0$ for all the electric length θ.

The return losses are characterized by a VSWR $S = 1 + \dfrac{\Delta R}{R_0}$:

$$Z_L = R_0 \frac{S + j tg\theta}{1 + jS tg\theta} = R_0 \frac{S(1 + tg^2\theta) + j(1 - S^2) tg\theta}{1 + S^2 tg^2\theta}$$

4) The variation of the load is written as:

$$Z_L : R_0 \rightarrow R_L + jX_L$$

This gives:

$$\begin{cases} \Delta R_L = R_L - R_0 = R_0(S - 1)\dfrac{1 - S tg^2\theta}{1 + S^2 tg^2\theta} \\[3mm] \Delta X_L = X_L = R_0(1 - S^2)\dfrac{tg\theta}{1 + S^2 tg^2\theta} \end{cases}$$

and the frequency variation is given by:

$$\Delta\omega = B(S - 1)\frac{-\alpha S tg^2\theta + (S + 1) tg\theta + \alpha}{1 + S^2 tg\,\theta}$$

With the two constants B and α, we get:

$$B = \frac{\dfrac{\partial R_T}{\partial I} R_0}{\dfrac{\partial R_T}{\partial I} \dfrac{\partial X_T}{\partial \omega} - \dfrac{\partial R_T}{\partial \omega} \dfrac{\partial X_T}{\partial I}}, \qquad \alpha = \frac{\partial X_T / \partial I}{\partial R_T / \partial I}$$

5) By putting $x = tg\theta$, we can see that this frequency variation is maximum when $\dfrac{\partial \Delta\omega}{\partial x} = 0$. This gives:

$$S^2 x^2 + 2\alpha S\,x - 1 = 0$$

With two solutions, we get:

$$\begin{cases} x_1 = tg\theta_1 = \dfrac{-\alpha - \sqrt{\alpha^2 + 1}}{S} \\[2mm] x_2 = tg\theta_2 = \dfrac{-\alpha + \sqrt{\alpha^2 + 1}}{S} \end{cases}$$

And after a long computation, we have the maximum excursion of frequency $\Delta\omega_2 - \Delta\omega_1$:

$$\Delta\omega_1 - \Delta\omega_2 = B\frac{S^2 - 1}{S}\sqrt{\alpha^2 + 1}$$

6) What is the sensibility?

If $\dfrac{\partial R_T}{\partial\omega} = 0$, then we have $B = \dfrac{R_0}{\dfrac{\partial X_T}{\partial\omega}}$ and with $Q_{ext} = \dfrac{\omega_0}{2R_0}\dfrac{\partial X_T}{\partial\omega}$,

we get:

$$\Delta\omega_{total} = \frac{\omega_0}{2Q_{ext}}\frac{S^2 - 1}{S}\sqrt{\alpha^2 + 1}$$

If more $\dfrac{\partial X_T}{\partial I} = 0$, then $\alpha = 0$ and we get:

$$\Delta\omega_{total} = \frac{\omega_0}{2Q_{ext}}\frac{S^2 - 1}{S}$$

which gives the sensibility of the oscillator:

$$\frac{\Delta f}{f_0} = \frac{S - \dfrac{1}{S}}{2Q_{ext}}$$

Then, variation of frequency due to the "pulling" Δf is slight if the circuit is good (Q_{ext} important).

16.5. Equivalent circuit of a DR coupled to a line

We want to realize a microwave oscillator with a DR and FET. First, we study the coupling of the DR with a microstrip line of characteristic impedance Z_0 (Figure 16.9).

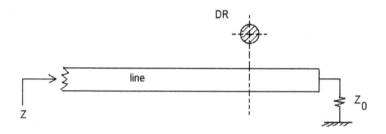

Figure 16.9. *The DR coupled with a line*

The DR with the line can be considered as a resonant circuit (R_r, C_r, L_r) with a magnetic coupling (L_m) to the line considered as induction L_l (Figure 16.10).

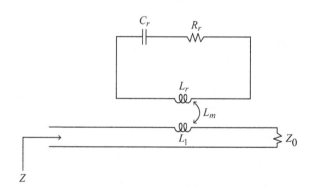

Figure 16.10. *The equivalent circuit*

1) Show that the input impedance can be written as:

$$Z = j\omega L_l + \cfrac{Q_0 \omega L_m^2}{L_r\left(\cfrac{\omega}{\omega_0} + jQ_0\left(1 - \cfrac{\omega_0^2}{\omega^2}\right)\right)}$$

with

$$\omega_0^2 L_r C_r = 1$$

and

$$R_r Q_0 = L_r \omega_0$$

2) For small values of induction of the line $(L_l = 0)$ and near the resonance $(\omega \cong \omega_0)$ show, this impedance can be written as:

$$Z \cong \cfrac{A}{1 + 2jQ_0 \cfrac{\omega - \omega_0}{\omega}}$$

Then, find the constant A as a function of the quantities L_m, L_r, Q_0 and ω.

3) Z' is the impedance of a parallel circuit R, L, C (Figure 16.11) with a frequency resonance $\omega = 1/\sqrt{LC}$. Show that if ω is near ω_0, we have the relation:

$$Z' \cong \cfrac{R}{1 + 2j\cfrac{R}{L\omega_0} \cdot \cfrac{\omega - \omega_0}{\omega}}$$

4) What are the conditions of identification of Z and near ω_0. To identify the resistance, we will consider $R(\omega) = R(\omega_0)$.

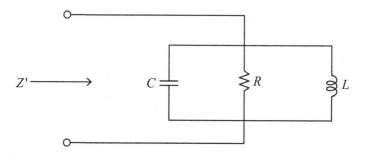

Figure 16.11. *Z' impedance*

5) Give rapidly a method to realize a microwave oscillator using a DR and FET.

The solution

1) The circuit in Figure 16.2 is redrawn as shown in Figure 16.12 and we have:

$$\begin{cases} v_1 = j\omega L_l i_1 + j\omega L_m i_2 \\ v_2 = j\omega L_m i_1 + j\omega L_r i_2 \end{cases}$$

But we have a load at the output:

$$v_2 = -\left(\frac{1}{j\omega C_r} + R_r \right) i_2$$

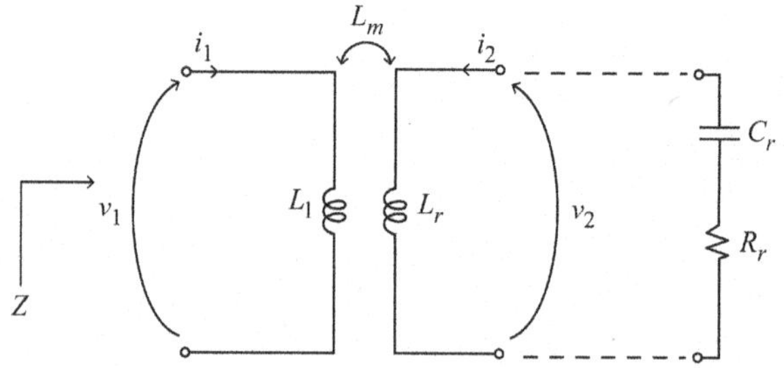

Figure 16.12. *The modified equivalent circuit*

Using the two equations, we get:

$$Z = \frac{v_1}{i_1} = j\omega L_l + \frac{\omega^2 L_m^2}{R_r + j\omega L_r + \dfrac{1}{j\omega C_r}}$$

and with the relations:

$$\omega_0^2 L_r C_r = 1 \text{ and } Q_0 R_r = L_r \omega_0$$

We have:

$$Z = j\omega L_l + \frac{Q_0 \omega L_m^2}{L_r \left[\dfrac{\omega}{\omega_0} + jQ_0 \left(1 - \dfrac{\omega_0^2}{\omega_0} \right) \right]}$$

2) For small values of induction $(L_l = 0)$ and near the resonance:

$$\begin{cases} \dfrac{\omega}{\omega_0} \approx 1 \\[2mm] 1 - \dfrac{\omega_0^2}{\omega^2} \approx 2\dfrac{\omega - \omega_0}{\omega} \end{cases}$$

The quantity Z becomes:

$$Z = \frac{Q_0 \omega L_m^2 / L_r}{1 + j2Q_0 \dfrac{\omega - \omega_0}{\omega}} = \frac{A}{1 + j2Q_0 \dfrac{\omega - \omega_0}{\omega}}$$

and A is a constant because when $\omega \approx \omega_0$:

$$A = \frac{Q_0 \omega L_m^2}{L_r} \approx \frac{Q_0 \omega_0 L_m^2}{L_r} \approx Q_0^2 \frac{L_m^2}{L_r^2} R_r$$

3) From Figure 16.11, we have:

$$Z' = \frac{L\omega}{\dfrac{L\omega}{R} - j\left(1 - \dfrac{\omega^2}{\omega_0^2}\right)}$$

But,

$$1 - \frac{\omega^2}{\omega_0^2} \approx -2\frac{\omega - \omega_0}{\omega_0}$$

and

$$Z' = \frac{R}{1 + 2j\dfrac{R}{L\omega_0}\dfrac{\omega - \omega_0}{\omega}}$$

4) The conditions of identification of Z and near ω_0 are then:

$$\begin{cases} R = \omega_0 Q_0 \dfrac{L_m^2}{L_r} = Q_0^2 \left(\dfrac{L_m^2}{L_r^2}\right) R_r \\[3mm] L = \left(\dfrac{L_m}{L_r}\right) L_m \\[3mm] C = \left(\dfrac{L_r^2}{L_m^2}\right) C_r \end{cases}$$

We verify that the new values of R, L, C have the dimensions of resistance, inductance and capacity.

5) Methods to realize a microwave oscillator using a DR and FET are mentioned in Chapter 15.

16.6. Bibliography

[BAH 03] BAHL I.J., BARTHIA P., *Microwave Solid State Circuit Design*, Wiley, 2003.

[GEN 84] GENTILI C., *Amplificateurs et Oscillateurs Microondes*, Masson, 1984.

[JAR 04] JARRY P., Circuits Actifs Microondes: Amplificateurs, Oscillateurs, University of Bordeaux, 2004.

Index

A, B, C

active element, 118, 179, 190
adaptation, 4, 137, 138
amplifier, 118, 125, 126, 131,
  141, 143, 153, 159, 160, 161,
  167
bandwidth, 29
broadband, 157
Butterworth, 86, 87, 103
capacitance, 219
cavity, 62, 65–67, 69, 89, 207,
  209
chain, 40, 63–66, 98–100, 111,
  162
characterization, 73, 74, 214, 215,
  228, 229
circles constant gain, 126–128
computing, 40, 79, 94, 133, 148,
  232, 235
configuration, 214–215
continuity, 77
coupled line, 3, 18, 105, 106, 109
coupling coefficient, 12, 25, 28,
  45, 47, 53, 209

D, E, F

determination, 37, 38, 45, 215,
  219–221
dielectric resonator (DR), 207
differential, 9, 10
direct coupling, 61, 83, 102
energy, 6, 7, 24–28, 48, 51, 67,
  74, 189, 191, 208, 209
equivalent circuit, 7, 8, 64–66, 73,
  81, 83, 108–110, 118, 182, 194,
  209, 210, 233, 244, 246
evanescent, 75–77
even mode, 11, 13, 14, 17, 19, 28,
  47, 48
factor
  noise, 118, 143, 147–150, 171
  quality, 208, 214
  Rollet, 133, 168
FET amplifier, 117
filter, 61, 69, 81, 83, 102, 103,
  105–107, 109
flow graph, 120, 123, 154

G, H

gain, 40–42, 117, 141–143, 153,
    154, 157, 160–163, 168, 169
gate, 118, 220
hybrid, 5, 33, 54–56

I, L, M

invertors, 91, 93, 94
limit, 133, 134, 198, 234, 235
linear, 118, 179, 187–189, 228,
    229, 232
loop, 41, 122–124, 154, 155
lossless, 7, 36, 37, 62, 65, 74, 97,
    99
low noise, 167
magic T, 33, 56
Masson's rule, 40, 122
mesh, 7, 179
microstrip, 3, 11, 95, 208, 244
modes, 11, 12, 14, 17–19, 26, 47,
    50, 74–77
modulation, 202–204

N, O

narrow band, 88, 103, 159, 202
noise
    circle, 149, 150
    factor, 118, 143, 147–150, 171
non-touching, 122, 154
non-unilateral, 128, 153
nonlinear, 178, 228, 229, 232
notch, 8, 40, 120–122
odd mode, 11, 12, 14, 17–19, 47,
    49, 50
order, 9, 10, 122–124, 141, 154,
    155, 160, 183

oscillation, 181–184, 193, 195,
    198, 208, 214, 221, 228–230,
    232, 239
overvoltage, 177, 209

P, Q, R

parameter,118, 131, 148, 149,
    157, 159, 167, 168, 177, 215,
    217, 220, 225, 228
pass-band, 86–88, 92
permittivity, 11
propagating, 11, 75, 77
pulling, 193, 239, 244
quasi-static, 177
reactance, 83, 89–92, 94, 102, 104
realization, 83, 91, 93, 214, 219
reflection,19–21, 28, 33, 37, 38,
    47, 54, 55, 62, 67, 74, 98, 101,
    120, 123, 131, 132, 153, 168,
    170, 209, 211, 212, 215, 217,
    219, 228–230, 232
resistance, 146, 149, 168, 171,
    178, 179, 181, 189, 190, 201,
    245, 248
resonance, 66, 67, 90, 209, 245,
    247
resonator, 61, 89, 207
response
    Butterworth, 87
    Tchebycheff, 87–88

S

scattering S, 29, 30, 34, 36, 38,
    39, 53, 62–64, 66, 97, 98, 118,
    131, 157, 159, 167, 168, 208,
    211, 217, 225, 228, 229, 232
simulation, 89, 91, 92
spectra, 177, 193

stability, 118, 131, 157, 168, 183,
  214, 219, 232, 237, 238
stage, 29, 47, 48, 50, 52
strip, 17
symmetric, 17, 34, 62, 64, 78, 99,
  111, 125, 153, 156
synchronization, 193, 233–235,
  237, 239
synthesis, 69, 83, 102, 103, 105,
  109, 112, 164, 165

transducer power gain, 123, 124
transmission, 19–21, 28, 47, 62,
  98, 101, 183, 209, 211, 212
two-poles, 228
unconditional, 131–133, 137, 140,
  141, 159, 161, 169
unilateral, 125, 126, 157, 160–163
wave, 14, 23, 39, 54, 63, 65–69,
  75, 77, 78, 90, 91, 94, 104, 202

T, U, W

Tchebycheff, 86, 87
Tee, 64

Printed in the United States
By Bookmasters